Lecture Notes in Economics and Mathematical Systems

295

Helmut Meister

The Purification Problem for Constrained Games with Incomplete Information

Springer-Verlag
Berlin Heidelberg New York London Paris Tokyo

Author

Dr. Helmut Meister
Fachbereich Mathematik und Informatik, Fernuniversität Hagen
Postfach 940, D-5800 Hagen 1, FRG

ISBN 3-540-18429-5 Springer-Verlag Berlin Heidelberg New York
ISBN 0-387-18429-5 Springer-Verlag New York Berlin Heidelberg

Library of Congress Cataloging-in-Publication Data. Meister, Helmut. The purification problem for constrained games with incomplete information. (Lecture notes in economics and mathematical systems; 295) Bibliography: p. 1. Game theory. 2. Statistical decision. I. Title. II. Series.
QA269.M45 1987 519.3 87-26520
ISBN 0-387-18429-5 (U.S.)

Printing and binding: Druckhaus Beltz, Hemsbach/Bergstr.
2142/3140-543210

CONTENTS

INTRODUCTION

The model of a noncooperative game as developed by J. Nash has been refined in the last decades mainly in two directions. First, games with constraints on the sets of decisions allowed to the players have been considered, whereby these constraints depend on the strategic behaviour of all players. This generalization is due to Debreu (1952). Later, games have been investigated, at which preferences of players are no longer laid down by payoff-functions, but are given by preference correspondences. Such an approach has for instance been chosen in a contribution of Shafer and Sonnenschein (1974). Second, recently, the study of games has been committed, at which decisions of players depend on informations. The informations are assumed to be different for different players. Games of this type have been investigated by Radner and Rosenthal (1982) as well as Aumann et al. (1983). The informations afford an insight into the state of the system, within which players have to make their decisions. Since these games are cut to situations, when there are uncertainties about the state of the system, they are called games with incomplete information. As players do not cooperate, they will hide their informations. Hence, there will be some unknown parameters of the state of the system for every player. Nevertheless, it will be assumed that there are some common beliefs on the variables determining the state of the system. These beliefs are represented by a probability distribution on these parameters which is assumed to be known to all players. This type of decision problem is, of course, essentially influenced by Bayesian decision theory.

Since modelling of decision situations often requires to take both aspects, constraints as well as uncertainties, into consideration, it seems to be appropriate to investigate such a model. Of course, constraints will be of some special feature in connection with games with incomplete information. The main interest besides existence of an equilibrium in randomized strategies will be the problem in how far pure strategies exist simulating randomized strategies as good as possible. Different questions arising in this context will be condensed by the concept of the purification problem.

Section 1.1 concerns the introduction of the model. It is a matter of a game with incomplete information, at which preferences of players are given by preference representation functions, and at which the set of allowed decisions for each player is laid down by constraint correspondences depending on the information of the player and the strategies of all players. The concept of a Nash equilibrium point is transferred to this model.

In Section 1.2 it will be explained what has to be understood by the purification problem, and it will be discussed in which sense the problem can be brought up to a solution.

First results in this context are presented in Section 1.3 . An essential resource is a purification lemma which allows to replace randomized strategies by payoff-equivalent pure strategies. Basing on this lemma, we will give an elegant proof for the denseness of the set of pure strategies w.r.t. the set of randomized strategies in case of a nonatomic information distribution.

Section 1.4 deals with sets of strategies concentrated on a correspondence. The behaviour of such sets will be investigated in case of disturbation of the correspondence. Moreover, the question, for which payoff-functions the expected payoff depends continuously on the strategy, will be broached.

Some existence-theorems for an equilibrium and an approximate equilibrium in pure strategies, respectively, are compiled in Section 1.5 . In particular, these theorems imply some existence results, when preferences of players are given by conventional payoff-functions.

In Section 1.6 , the question of pure-strategy equilibrium existence is raised. It turns out that this problem has a satisfactory solution in case of a certain payoff-structure and some special type of constraints.

Finally, Section 1.7 gives some hints, how approximately payoff-equivalent pure strategies can be determined, when the information space is discrete. Moreover, it will be shown that most information spaces allow a suitable discretization.

In § 2, the results are applied to a market game as developed by Hildenbrand (1974), for instance. In Section 2.1 it will be shown that the market game can be arranged in the theory of constrained games with incomplete information. To this end, we introduce the concept of a randomized allocation. With the aid of the existence results as derived in 1.5 the existence of a Walras-allocation will be proved.

In Section 2.2 , the concept of core will be established for randomized allocations. Every randomized Walras-allocation proves to be an element of this extended core.

A standard problem from statistical decision theory is treated in § 3 .

Section 3.1 deals with minimax decisions. Primary, the relation to the theory of games with incomplete information will be emphasized.

Finally, in Section 3.2 , we consider set-valued estimators. Two different approaches will be presented both allowing to establish the concept of a set-valued minimax estimator. Set-valued estimators prove to be pure strategies for the decision problems in question. This fact suggests to apply the results of § 1 to show existence of approximate set-valued minimax estimators.

Extracts from literature, to which this treatise refers, are summarized in the appendix.

Finally, I would like to acknowledge my indeptedness to Prof. O. Moeschlin for a lot of interesting discussions and valuable suggestions concerning especially Section 3.2. I am grateful to Mrs. G. Soentgen who has type-written the manuscript in a very careful and competent way.

NOTATIONS

We introduce some symbols and some notations:

$\mathbb{R}$ denotes the set of real numbers,

$\overline{\mathbb{R}}$ denotes the set of extended real numbers,

$\mathbb{R}_+$ denotes the set of nonnegative real numbers,

$\mathbb{N}$ denotes the set of natural numbers,

$\mathbb{C}$ denotes the set of complex numbers.

For $z \in \mathbb{C} - \{0\}$ the abbreviation $\arg z$ means the argument of z, i.e. the unique angle $\varphi \in [0,2\pi)$ from the polar coordinate representation $z = re^{i\varphi}$ with $r > 0$.

For $x \in \mathbb{R}^n$ we mean by x_i the i-th component of x. This will usually not be specially emphasized. We assume $x = (x_1,...,x_n)$ to be a line vector, respectively. The transposed vector of x is denoted by x^t.

The limes inferior of a sequence (x_n) from $\mathbb{R}$ is denoted by $\underline{\lim}\, x_n := \lim\inf x_n$. Accordingly, for the limes superior we use the symbol $\overline{\lim}\, x_n := \lim\sup x_n$.

The same kind of notation is used for the limes inferior and limes superior, respectively, of a sequence of sets.

Let S be any set. Instead of the phrase "for every $x \in S$" we simply write "$(x \in S)$". The power set of S is denoted by $\mathcal{P}(S)$. For every topological space S, the (Borel) σ-field generated by the topology of S is denoted by $\mathcal{B}(S)$. For the system of compact nonvoid subsets of S, we use the symbol $\mathcal{C}(S)$. Some properties of this space are summerized in A 2. For every set $A \subset S$ let $\operatorname{int} A$ be the set of interior points of A, i.e.

$$\operatorname{int} A := \bigcup_{\substack{U \subset A \\ U \text{ open}}} U .$$

Let V be a real vector space, and let A be a subset of V. By conv A, we mean the convex hull of A, i.e.

$$\operatorname{conv} A := \{x \in V \mid \exists n \in \mathbb{N},\ \exists x_1,\dots,x_n \in A, \\ \exists \alpha_1,\dots,\alpha_n \geq 0 : \sum \alpha_i = 1 \\ \text{and } x = \sum_{i=1}^{n} \alpha_i x_i\} .$$

Let Ω and Ω' be both nonempty. By a <u>correspondence</u> from Ω to Ω' we mean a set-valued mapping $F : \Omega \to \mathcal{P}(\Omega')$ with $F(\omega) \neq \emptyset$ $(\omega \in \Omega)$.

The <u>graph</u> of F is defined by

$$\operatorname{graph} F := \{(\omega,\omega') \in \Omega\times\Omega' \mid \omega' \in F(\omega)\} .$$

Let $(\Omega,\mathcal{A})$ be a measurable space. The <u>characteristic function</u> of an arbitrary set $A \in \mathcal{A}$ is denoted by 1_A. For every $\omega \in \Omega$ let ε_ω be the <u>point measure</u> at the point ω, i.e.

$$\varepsilon_\omega(A) := 1_A(\omega) \qquad (A \in \mathcal{A}) .$$

Let Ω be additionally a topolological space, let $\mathcal{A} := \mathcal{B}(\Omega)$ and let P be a probability measure on $\mathcal{A}$. Then, the set

$$\operatorname{spt} P := \bigcap_{\substack{F \subset \Omega,\, P(F)=1 \\ F \text{ closed}}} F$$

is called the <u>support</u> of P (obviously, spt P is the smallest closed set of measure 1). Whenever the topology of Ω has a countable base, we have

$$P(\operatorname{spt} P) = 1 .$$

§ 1 THE PURIFICATION PROBLEM IN THE GAME-THEORETIC CONTEXT

1.1 A constrained game with incomplete information

The following noncooperative game is a more general version of the usual model of a noncooperative game developped by J. Nash in 1950. The model considered here generalizes the model of Nash in two respects: First, players choose their acts dependent on private informations, and second, there are constraint correspondences which players have to observe when choosing their acts. The strategies of the players are decision functions, which attach an act to each information. The information a player obtains will thereby in general not be equal to the whole information available, and there will be no chance for a direct inference on the information of the other players. For that reason games of this type are called games with incomplete information.

We assume a finite number of participants. Let $I := \{1,\dots,N\}$ be the set of players.

We assume further that some parameters of the game are not completely known; i.e., there are some parameters which are not observable for all players. But, each player can observe one of these parameters. The space of observable parameters of player i is denoted by Ω_i which is assumed to be a separable metric space and is endowed with the corresponding Borel σ-field $A_i := \mathcal{B}(\Omega_i)$. Further, we assume that there are some common beliefs of all players on the unknown parameters. These beliefs are represented by a finite measure μ on $\bigotimes_{i\in I} \mathcal{B}(\Omega_i)$, which is assumed to be known to all players.

The marginal measure μ_i of μ on Ω_i is the information distribution of player i. The measure space (Ω_i, A_i, μ_i) is called the <u>information space of player i</u>.

The information space (Ω_i, A_i, μ_i) is called

- <u>Polish</u>, if Ω_i is a Polish space, i.e. if Ω_i is additionally complete;
- <u>locally compact</u>, if Ω_i is a locally compact space;
- <u>nonatomic</u>, if μ_i is a nonatomic measure on A_i.

After observing his information variable each player i chooses an act from his decision space S_i, which is assumed to be Polish and is endowed with its Borel σ-field $\mathcal{S}_i = \mathcal{B}(S_i)$. The measure space $(S_i, \mathcal{S}_i)$ is called the <u>decision space of player i</u>. The decision space $(S_i, \mathcal{S}_i)$ is called <u>compact</u>, if S_i is a compact (metric) space.

In order to have some kind of randomization for strategies, we will introduce a <u>(mixed) strategy of player i</u> by a stochastic kernel $K_i : \Omega_i \times \mathcal{S}_i \to [0,1]$, i.e. by a function K_i for which

- $K_i(\omega_i, .)$ is a probability measure on $\mathcal{S}_i$ for all $\omega_i \in \Omega_i$;
- $K_i(., A_i)$ is a measurable function on Ω_i for all $A_i \in \mathcal{S}_i$.

For all strategies K_i let $\mu_i \otimes K_i$ be the measure on $A_i \otimes \mathcal{S}_i$ defined by the measure μ_i and the stochastic kernel K_i. We identify two strategies K_i and K_i', when they coincide μ_i-a.e., or equivalently, when $\mu_i \otimes K_i = \mu_i \otimes K_i'$.

A pure strategy of player i is a mixed strategy K_i for which $K_i(\omega_i,.)$ is a point mass μ_i-a.e. Thus, a pure strategy may be identified with a measurable function $f_i : \Omega_i \to S_i$. The stochastic kernel associated with $f_i : \Omega_i \to S_i$ will be denoted by ε_{f_i} and it satisfies

$$\varepsilon_{f_i}(\omega_i, A_i) = 1_{A_i}(f_i(\omega_i)) \qquad (\omega_i \in \Omega_i, A_i \in S_i).$$

Let Σ_i be the set of all mixed strategies of player i. Then, Σ_i is a convex set.

Since Ω_i and S_i are both (separable) metric spaces, the strategy space Σ_i can be endowed with the corresponding weak topology. A sequence $(K_i^{(n)} | n \in \mathbb{N})$ from Σ_i is said to converge to $K_i \in \Sigma_i$, if and only if

$$\lim_{n\to\infty} \int h \, d\mu_i \otimes K_i^{(n)} = \int h \, d\mu_i \otimes K_i$$

for every bounded continuous function $h : \Omega_i \times S_i \to \mathbb{R}$. The weak topology on Σ_i is metrizable (cf. A. 1.3) . The space

$$\Sigma := \mathop{\times}_{i=1}^{n} \Sigma_i$$

endowed with the corresponding product topology is therefore a metric space, too.

In order to accommodate as large variety of application as possible, we introduce constraint correspondences for all players. The meaning of these correspondences is that they restrict the set of allowed decisions of each player dependent on his information and the stratetical behaviour of all players. The constraint correspondence of player i is therefore a function

$$R_i : \Omega_i \times \Sigma \to S_i - \{\emptyset\} .$$

Finally, the behaviour of player i is assumed to be influenced by the <u>preference representation function</u> (PRF) (of player i)

$$v_i : \Sigma \times \Sigma_i \to \mathbb{R} ,$$

which has the following properties

- $v_i(K, K_i) = 0 \quad (K = (K_1, \ldots, K_N) \in \Sigma)$
- $v_i(K, .)$ is quasiconcave on Σ_i for given $K \in \Sigma$, i.e., the set $\{L_i \in \Sigma_i | v_i(K, L_i) > \alpha\}$ is convex for every $\alpha \in \mathbb{R}$.

The value $v_i(K, L_i)$ quotes the profit of player i, when he replaces his strategy K_i in the strategy N-tuple $(K_1, \ldots, K_N)$ by the new strategy L_i.

The following examples illustrate the concept of a PRF and a constraint correspondence, resp.

<u>1.1.1 Examples:</u>

(1) Suppose, every player $i \in I$ is given a continuous bounded payoff function

$$\mu_i : \bigtimes_{j=1}^{N} \Omega_j \times \bigtimes_{j=1}^{N} S_j \to \mathbb{R} ,$$

and suppose, the informations of players are independent. Then the corresponding PRF v_i of player i is defined by

$$v_i(K, L_i) := \int \mu_i \, dQ(K|L_i) - \int \mu_i \, dQ(K|K_i) ,$$

where

$$Q(K|K_i') := \left(\bigotimes_{j=1}^{i-1} \mu_j \otimes K_j \right) \otimes (\mu_i \otimes K_i') \otimes \bigotimes_{j=i+1}^{N} \mu_j \otimes K_j$$

for every $K = (K_1,\dots,K_N) \in \Sigma$ and $K_i' \in \Sigma_i$.
In this case, the PRF v_i quotes the change in the expected payoff, when player i replaces his strategy K_i by strategy L_i.

(2) Let be given a payoff function u_i for each player i as in the first example and let

$$w_i(\omega_i,s_i,K) := \int u_i(\omega_1,\dots,\omega_n,s_1,\dots,s_n)\, d \bigotimes_{j \neq i} \mu_j \otimes K_j$$

$$(\omega_i \in \Omega_i, s_i \in S_i; K=(K_1,\dots K_N) \in \Sigma) .$$

If player i accepts only optimal decisions w.r.t. w_i, an appropriate constraint correspondence for him will be given by

$$R_i(\omega_i,K) := \{s_i \in S_i | w_i(\omega_i,s_i,K) = \max_{t_i \in S_i} w_i(\omega_i,t_i,K)\} .$$

(3) Let the game for simplicity be played by one single player. Let (Ω,A,μ) be his information space; and let $(S,\mathcal{S})$ be his decision space. Further, let $u:\Omega\times S \to \mathbb{R}^n$ be a bounded continuous (vector-valued) function quoting the profit dependent on information and decision of the player under n different aspects. The player's preferences on the set $\mathbb{R}^n$ of all payoff vectors coming into question are assumed to be given by a correspondence

$$P:\mathbb{R}^n \to \mathcal{P}(\mathbb{R}^n) .$$

The set $P(x)$ is the set of all payoff configurations preferred to $x \in \mathbb{R}^n$. In case $P(x)$ is convex for all $x \in \mathbb{R}^n$ and the graph of P is open, one can show that a continuous function $w:\mathbb{R}^n \times \mathbb{R}^n \to \mathbb{R}_+$ with the property

$$w(x,y) > 0 \iff y \in P(x) \quad (x,y \in \mathbb{R}^n)$$

exists, such that $w(x,.)$ is quasiconcave for every $x \in \mathbb{R}^n$. (As in Shafer, Sonnenschein (1975) one may choose for $w(x,y)$ the distance between (x,y) and the complement of the graph of P.) In this situation an appropriate PRF v is defined by

$$v(K,L) := w\left(\int u \, d\mu\otimes K , \int u \, d\mu\otimes L\right) \qquad (K,L \in \Sigma) .$$

The described model of a game defined by the 7N-tuple

$$(\Omega_i, A_i, \mu_i; S_i, \mathcal{S}_i; R_i, v_i \mid i \in I)$$

will be called a <u>game with incomplete information</u>. The solution concept of a Nash equilibrium point can be transferred to this game.

<u>1.1.2 Definition:</u> *Let* $K := (K_1,\ldots,K_N) \in \Sigma$.

(1) *A strategy* L_i *of player i is called an* <u>*admissible strategy w.r.t.* K *(of player i)*</u>, *iff*

$$L_i(\omega_i, R_i(\omega_i, K)) = 1 \quad \mu\text{-a.e.}$$

holds.

(2) *A strategy N-tuple* K *is called a* <u>*(Nash-) equilibrium point*</u> *of the game, iff* K_i *is an admissible strategy w.r.t.* K *of every player i, and*

$$v_i(K, L_i) \leq 0$$

holds for all admissible strategies $L_i \in \Sigma_i$ *w.r.t.* K.

Games of this type with different specializations have been studied by Milgrom and Weber (1980), Radner and Rosenthal (1982) as well as Aumann et al. (1983). All authors consider games for which preferences are given by conventional payoff-functions.

The last two contributions are confined to finite decision spaces. Constraint correspondences are investigated in none of the above contributions. Although games with constraints and PRFs are studied by Wieczorek (1984) in a general set-up, there is no direct relation to the specific questions appearing in connection with games with incomplete information except the question of equilibrium existence.

Problems arising in connection with the lack of an information distribution known to all players - as assumed in the game described here - are discussed in a paper of Mertens and Zamir (1985).

It should be mentioned that the information space (Ω_i, A_i, μ_i) can have a different interpretation in some mathematical models such as market games. In a market game the space Ω_i means a group of players, the σ-field A_i means the system of possible coalitions and the measure μ_i gives the share of every coalition w.r.t. the set of all participants. Nevertheless, in § 2 will be shown in how far market games may also be considered as games with incomplete information.

If Ω_i is chosen as a one-point set, even such cases can be accommodated to the model, when player i cannot observe any information. If R_i is defined by

$$R_i(\omega_i, K) := S_i \quad (\omega_i \in \Omega_i, K \in \Sigma) ,$$

the constraint correspondence of player i is ineffective. By this means the model includes the situation, when some players have no constraints. Especially the noncooperative game introduced by Nash is therefore a game of the type considered here.

The noncooperative character of the game becomes apparent by the fact, that players don't tell their informations to the others. But there may be, of course, some weak form of cooperation, since the constraints may arise from arrangements between players.

1.2 The purification problem

In the game described in 1.1 the strategies of player i have been stochastic kernels from the information space (Ω_i, A_i, μ_i) into the decision space $(S_i, \mathcal{S}_i)$. Since decisions depend on informations, one obtains even in case of pure strategies a certain kind of mixture of decisions. This fact gives rise to the question, in which situation players can be satisfied to use pure strategies without changing the game essentially.

The mathematical methods available for the proof of existence of an equilibrium for the game described in 1.1 will require the convexity of the set of strategies, and therefore some kind of randomization. Nevertheless, in most situations appearing in practice players will use pure strategies, not only for the reason of easier handling, rather because they prove to be equivalent to mixed strategies under several aspects. Some of those aspects being important will be listed in the sequel.

For each strategy $K_i \in \Sigma_i$ let the corresponding <u>marginal strategy</u> $(\mu_i \otimes K_i)_{S_i}$ on S_i be defined by

$$(\mu_i \otimes K_i)_{S_i} := \pi_{S_i}(\mu_i \otimes K_i),$$

where π_{S_i} is the projection of $\Omega_i \times S_i$ onto S_i. Each pure strategy $\varepsilon_{f_i} \in \Sigma_i$ satisfies

$$(\mu_i \otimes \varepsilon_{f_i})_{S_i} = f_i(\mu_i) .$$

The marginal strategy of a strategy of player i has a vivid

interpretation. An observer who cannot perceive the informations $\omega_i \in \Omega_i$, but realizes the strategic behaviour of player i on S_i, will describe the strategy by its marginal distribution on S_i. Even in case of a pure strategy, the marginal strategy is in general no point mass.

A pure strategy f_i is called <u>mixture preserving</u> w.r.t a mixed strategy K_i, iff

(M) $$(\mu_i \otimes K_i)_{S_i} = (\mu_i \otimes \varepsilon_{f_i})_{S_i}$$

is satisfied.

With regard to the given constraints, it will be important that the pure strategy f_i is concentrated on the same constraint correspondence as K_i, what can be ensured by the requirement

(C) $$f_i(\omega_i) \in \text{spt } K_i(\omega_i,.) \qquad \mu_i\text{-a.e.}$$

Every pure strategy satisfying (C) is called a <u>conservative</u> strategy (w.r.t. K_i).

Let be given a $\mu_i \otimes K_i$-integrable vector-valued payoff-function $\mu_i : \Omega_i \times S_i \to \mathbb{R}^n$. Then, every pure strategy ε_{f_i} is called <u>payoff-equivalent</u> (to K_i), iff μ_i is $\mu_i \otimes \varepsilon_{f_i}$-integrable, too, and

(E) $$\int \mu_i \, d\mu_i \otimes \varepsilon_{f_i} = \int \mu_i \, d\mu_i \otimes K_i$$

is satisfied.

The following example shows that the claims (M) and (C) as well as (M) and (E) are generally incompatible.

<u>1.2.1 Example:</u> We assume that there is only one player. Let his information space be given by $\Omega := [0,1]$ together with the corresponding Borel σ-field and the Lebesgue measure λ on $[0,1]$. His decision space is assumed to be $S := [0,2]$ together with the Borel σ-field on S. Further, let be given the strategy K with

$$K(\omega,.) := \frac{1}{2}\varepsilon_{\omega} + \frac{1}{2}\varepsilon_{\omega+1} \qquad (\omega\in\Omega)$$

and $u : \Omega\times S \to \mathbb{R}$ with

$$u(\omega,s) := \cos(2\pi(\omega-s)) \qquad (\omega\in\Omega, s\in S) .$$

For every pure strategy ε_f with

(C) $$f(\omega) \in \{\omega,\omega+1\} \quad \lambda\text{-a.e.}$$

we have

$$\lambda\{f\in B\} = \lambda(A\cap B) \qquad (B\in\mathcal{B}(S),\ B\subset[0,1]),$$

where

$$A := \{\omega\in\Omega \mid f(\omega) = \omega\} .$$

Since

$$(\lambda\otimes K)_S = \frac{1}{2}\lambda_{[0,2]},$$

the validity of

(M) $$(\lambda\otimes\varepsilon_f)_S = (\lambda\otimes K)_S$$

would imply

$$\lambda(A\cap B) = \lambda\{f\in B\} = (\lambda\otimes\varepsilon_f)_S(B) = (\lambda\otimes K)_S(B) =$$

$$= \frac{1}{2}\lambda(B) \qquad (B\in\mathcal{B}(S),\ B\subset[0,1]) .$$

Setting $B := A$, we conclude $\lambda(A)=0$, and setting $B := [0,1]$, we conclude $\lambda(A) = \frac{1}{2}$. This shows that (C) and (M) cannot be both satisfied.

Suppose that the equation

$$\text{(E)} \qquad \int u \, d\lambda\otimes\varepsilon_f = \int u \, d\lambda\otimes K$$

holds for the pure strategy ε_f. Then, as a consequence of the identities

$$\int u \, d\lambda\otimes K = \frac{1}{2}\cos 0 + \frac{1}{2}\cos 2\pi = 1$$

$$\int u \, d\lambda\otimes\varepsilon_f = \int_0^1 \cos(2\pi(\omega - f(\omega)))\, d\omega$$

we have

$$f(\omega) \in \{\omega, \omega+1\} \qquad \lambda\text{-a.e.}$$

and therefore (C). Hence, even (E) and (M) are incompatibel.

These negative results give rise to the question whether there is at least an approximate solution, when (C), (E) and (M) are claimed simultaneously. This problem will be considered in connection with the set-up of section 1.1 .

Let be given a PRF v_i for every player i. For given strategies $K_1,\ldots,K_N$, a corresponding family $\varepsilon_{f_1},\ldots,\varepsilon_{f_N}$ of pure strategies will be called a <u>purification</u> (of $K_1,\ldots,K_N$), iff

$$\text{(P)} \qquad v_i(\varepsilon_{f_1},\ldots,\varepsilon_{f_N},L_i) = v_i(K_1,\ldots,K_N,L_i) \qquad (L_i\in\Sigma_i;\ i\in I)$$

holds. This would ensure the existence of an equilibrium in pure strategies, whenever $(K_1,\ldots,K_N)$ was an equilibrium - refrained from constraints. We will later give an example showing that there exist mixed strategy equilibria without a purification even in very specialized situations. Nevertheless, since v_i is continuous on $\Sigma\otimes\Sigma_i$, claim (P) can in case of compactness of the strategy spaces Σ_j $(j\in I)$ always be satisfied approximately, provided that, given any strategy $K_i \in \Sigma_i$, there exists a sequence $(\varepsilon_{f_i^{(t)}} \mid t\in\mathbb{N})$ of pure strategies with

$$\text{(APPR)} \qquad \lim_{t\to\infty} \varepsilon_{f_i^{(t)}} = K_i \,.$$

Every sequence $(\varepsilon_{f_i^{(t)}} \mid t\in\mathbb{N})$ with the approximation property (APPR) consisting of conservative strategies, i.e. satisfying in addition

$$\text{(C)} \qquad f_i^{(t)}(\omega_i) \in \text{spt}\, K_i(\omega_i,.) \quad \mu_i\text{-a.e.}\,,$$

is called a <u>purifying sequence</u> for K_i or a <u>sequence of approximate purifications</u> for K_i. It seems to be advisible to look for purifying sequences, because every sequence of this type is asymptotically mixture preserving, i.e. we have

$$\lim_{t\to\infty} (\mu_i\otimes\varepsilon_{f_i^{(t)}})_{S_i} = (\mu_i\otimes K_i)_{S_i}$$

- the latter property is a consequence of the continuity of the projection π_{S_i}.

In Section 1.4 we will show that the mapping

$$K_i \longmapsto \int u_i \, d\mu_i\otimes K_i$$

is continuous for a comprehensive class of payoff-functions $u_i:\Omega_i\times S_i \to \mathbb{R}^n$. Every purifying sequence $(\varepsilon_{f_i^{(t)}})$ for K_i satisfies therefore

$$\lim_{t\to\infty} \int u_i \, d\mu_i\otimes\varepsilon_{f_i^{(t)}} = \int u_i \, d\mu_i\otimes K_i$$

for such payoff-functions u_i. Hence, the given sequence is approximately payoff-equivalent.

It should be mentioned that the claims (C), (E) and (M) can be satisfied in case of finite decision spaces S_i as it is shown by Radner and Rosenthal (1982).

The problem of existence of pure strategies satisfying (C), (E), (M), (P) and (APPR), resp., at least approximately, is called the <u>purification problem</u>. The following statements give partial solutions of this problem and show respectively in how far the problem considered here can be weakened in order to have a solution.

1.3 On existence of approximate purifications

A partial solution of the purification problem as introduced in section 1.2 will now be given. We will derive an existence theorem for payoff-equivalent pure strategies w.r.t. a vector valued payoff function. The theorem is the starting-point for the proof of existence of purifying sequences.

In so far as not otherwise arranged, let be given an information space (Ω, A, μ) and a Polish decision space $(S, \mathcal{S})$. Further, let Σ be the set of all strategies $K: \Omega \times \mathcal{S} \to [0,1]$ and let $\mathcal{L}^n(\Sigma)$ be the set of all functions $u: \Omega \times S \to \mathbb{R}^n_+$, which are $\mu \otimes K$-integrable w.r.t. every strategy $K \in \Sigma$. We will first give a characterization of the set of all these functions.

1.3.1 Lemma: *For each measurable function* $u: \Omega \times S \to \mathbb{R}^n_+$ *the following statements are equivalent:*

(1) $u \in \mathcal{L}^n(\Sigma)$,

(2) $\sup_{s \in S} u_i(\cdot, s)$ is a μ-integrable function,

(3) there exists a μ-integrable function g such that

$$\max_{1 \leq i \leq n} u_i(\omega, s) \leq g(\omega) \quad (s \in S) \quad \mu\text{-a.e.}$$

1) *According to A.1.10* $\sup_{s \in S} u_i(\cdot, s)$ *is a* μ*-a.e. measurable function.*

Proof:

(1) $\Longrightarrow$ (2): Suppose,

$$\int \sup_{s\in S} u_i(.,s)\, d\mu = \infty$$

is satisfied for some $i \in \{1,...,n\}$. Using standard methods, one can construct a sequence (Z_n) of nonnegative measurable functions $z_n: \Omega \to \mathbb{R}_+$ with the following properties:

- $$z_n \uparrow \sup_{s\in S} u_i(.,s) \quad \mu\text{-a.e.}$$

- $$\int z_n\, d\mu \geq 2^n$$

- $$F_n(\omega) := \{s\in S \mid z_n(\omega) \leq u_i(\omega,s) \leq \sup_{s\in S} u_i(\omega,s)\} \neq \emptyset \quad (\omega\in\Omega) .$$

Since the graph of the correspondence F_n is measurable w.r.t. the σ-field $\hat{A}\otimes S$ ($\hat{A}$ denoting the μ-completion of A), there is a measurable selection f_n of F_n (cf. A.2.10). The corresponding strategy f_n satisfies

$$\int u_i \, d\mu\otimes\varepsilon_{f_n} = \int u_i(\omega, f_n(\omega))\, d\mu \geq \int Z_n\, d\mu \geq 2^n .$$

Setting

$$K(\omega,.) := \sum_{n\in\mathbb{N}} \frac{1}{2^n} \varepsilon_{f_n(\omega)} \qquad (\omega\in\Omega) ,$$

we obtain a strategy K with

$$\int u_i \, d\mu\otimes K = \sum_{n\in\mathbb{N}} \frac{1}{2^n} \int u_i \, d\mu\otimes\varepsilon_{f_n} \geq \sum_{n\in\mathbb{N}} 1 = \infty .$$

This calculation shows $u \notin L^n(\Sigma)$ and proves therefore the implication "(1) $\Longrightarrow$ (2)".

(2) $\Longrightarrow$ (3): Since $\sup_{s \in S} u_i(.,s)$ $(i=1,\ldots,n)$ is $\hat{A}$-measurable, there is an A-measurable function $g:\Omega \to \overline{\mathbb{R}}_+$ with the property

$$g = \sum_{i=1}^{n} \sup_{s \in S} u_i(.,s) \qquad \mu\text{-a.e.}$$

If assertion (2) holds, we have immediately that g is a function as required in (3).

(3) $\Longrightarrow$ (1): If (3) holds true for some μ-integrable function g, we have the estimation

$$\int u_i \, d\mu \otimes K \leq \int \sup_{s \in S} u_i(.,s) \, d\mu \leq \int g \, d\mu < \infty \qquad (K \in \Sigma;\ i=1,\ldots,n).$$

Hence, $u \in L^n(\Sigma)$. □

Before proving a first result on payoff-equivalence, we introduce a certain type of strategies. It is a matter of strategies concentrated on finitely many pure strategies.

<u>1.3.2 Definition:</u> *The strategy* K *is called <u>discrete</u>, iff* K *is of the type*

$$K(\omega,.) = \sum_{1 \leq i \leq r} \alpha_i(\omega)\, \varepsilon_{f_i(\omega)} \qquad \mu\text{-a.e.}$$

for some $r \in \mathbb{N}$ *and some measurable functions* $f_1,\ldots,f_r:\Omega \to S$ *as well as* $\alpha_1,\ldots,\alpha_r:\Omega \to [0,1]$ *with* $\sum_{1 \leq i \leq r} \alpha_i = 1$. *For a strategy* K *of this type we introduce the symbol*

$$K =: \delta_{(f_1,\ldots,f_r,\alpha_1,\ldots,\alpha_r)}.$$

Especially each pure strategy ε_f is a discrete strategy; we have $\varepsilon_f = \delta_{(f,1)}$. Obviously, the functions $f_1,\ldots,f_r$ and $\alpha_1,\ldots,\alpha_r$ in Definition 1.3.2 are not uniquely determined.

But this will prove to be not important for the further proceeding. The following lemma shows that each mixed strategy allows a payoff-equivalent discretization.

<u>1.3.3 Lemma:</u> *Let* $K \in \Sigma$ *and* $u \in L^n(\Sigma)$ *be given. Further, let* $F: \Omega \to S$ *be a correspondence with a measurable graph; and suppose*

$$K(\omega, F(\omega)) = 1 \quad \mu\text{-a.e.}$$

Then, there exists a discrete strategy $\delta := \delta_{(f_1, \ldots, f_r, \alpha_1, \ldots, \alpha_r)}$ *with the properties*

(1) $\int u(\omega,s)\, \delta(\omega,ds) = \int u(\omega,s)\, K(\omega,ds) \quad \mu\text{-a.e.}$

(2) $f_1(\omega), \ldots, f_r(\omega) \in F(\omega)$.

Proof: Define the set

$$C(\omega) := \operatorname{conv}\{u(\omega,s) \mid s \in S\} \qquad (\omega \in \Omega).$$

Then, A.3.4 implies

$$\int u(\omega,s)\, K(\omega,ds) \in C(\omega)$$

provided that $u(\omega,.)$ is a $K(\omega,.)$-integrable function. Therefore, observing A.3.1, there exist $s_1, \ldots, s_{n+1} \in S$ and $\alpha_1, \ldots, \alpha_{n+1} \geq 0$ with $\sum \alpha_k = 1$ s.t.

$$\int u(\omega,s)\, K(\omega,ds) = \sum_{k=1}^{n+1} \alpha_k\, u(\omega, s_k) .$$

If the correspondence G is defined by

$$G(\omega) := \{(s_1, \ldots, s_{n+1}, \alpha_1, \ldots, \alpha_{n+1}) \in S^{n+1} \times \mathbb{R}_+^{n+1} \mid \sum \alpha_k = 1, \int u(\omega,s)\, K(\omega,ds) = \sum_{k=1}^{n+1} \alpha_k\, u(\omega,s_k)\} \qquad (\omega \in \Omega),$$

the preliminary considerations show that $G(\omega) \neq \emptyset$ μ-a.e. Hence, by A.2.10, G allows a measurable selection , i.e., there are measurable functions $f_1,\dots,f_{n+1} : \Omega \to S$ and $\alpha_1,\dots,\alpha_{n+1} : \Omega \to \mathbb{R}_+$ satisfying

$$\sum_{k=1}^{n+1} \alpha_k(\omega) = 1, \quad \int u(\omega,s)\, K(\omega,ds) = \sum_{k=1}^{n+1} \alpha_k(\omega)\, u(\omega,f_k(\omega))$$

μ-a.e.

Define now the discrete strategy δ by

$$\delta(\omega,.) := \sum_{k=1}^{n+1} \alpha_k(\omega)\, \varepsilon_{f_k(\omega)} \qquad (\omega \in \Omega) .$$

Then, δ satisfies (1). Property (2) is an immediate consequence of (1), when u is supplied with the additional component $1_{\text{Graph}(F)}$. □

In case μ is nonatomic, we obtain a stronger version of Lemma 1.3.3 .

<u>1.3.4 Corollary</u>: *Let μ be a nonatomic measure. Further, let the assumptions be given as in 1.3.3 . Then, a pure strategy ε_f with properties*

(1) $$\int u \; d\mu \otimes \varepsilon_f = \int u \; d\mu \otimes K$$

(2) $$f(\omega) \in F(\omega) \quad \mu\text{-a.e.}$$

exists.

Proof: First, a discrete strategy $\delta_{(f_1,\dots,f_{n+1},\alpha_1,\dots,\alpha_{n+1})}$ with properties (1) and (2) in 1.3.3 exists. Consider now the finite measures μ_{jk} with density functions $u_j(.,f_k(.))$ w.r.t. μ $(j=1,\dots,n;\ k=1,\dots,n+1)$. Then, by A.3.2, we have a measurable partition $D_1,\dots,D_{n+1}$ of Ω s.t.

$$\int \alpha_k \, d\mu_{jk} = \mu_{jk}(D_k) \qquad (j=1,\dots,n;\ k=1,\dots,n+1).$$

The function $f: \Omega \to S$, defined by

$$f(\omega) := f_k(\omega) \quad (\omega \in D_k, \ k=1,\ldots,n+1)$$

is measurable and satisfies

$$\int u_j(\omega, f(\omega)) \, d\mu = \sum_{k=1}^{n+1} \int_{D_k} u_j(\omega, f_k(\omega)) \, d\mu =$$

$$= \sum_{k=1}^{n+1} \mu_{jk}(D_k) = \sum_{k=1}^{n+1} \int \alpha_k \, d\mu_{jk} =$$

$$= \sum_{k=1}^{n+1} \int \alpha_k(\omega) \, u_j(\omega, f_k(\omega)) \, d\mu =$$

$$= \int u_j \, d\mu \otimes \delta = \int u_j \, d\mu \otimes K \quad (j=1,\ldots,n) \; ;$$

the last equation following from 1.3.3 . Further, since $f_k(\omega) \in F(\omega)$ μ-a.e. $(k=1,\ldots,n+1)$ we conclude

$$f(\omega) \in F(\omega) \quad \mu\text{-a.e.}$$

Therefore, ε_f has the desired properties. □

As in 1.1 we equip the strategy set Σ with the topology of weak convergence. Since Ω and S are separable metric spaces, by A.1.2 , there exists a sequence $(h_k | k \in \mathbb{N})$ of bounded continuous functions $h_k : \Omega \times S \to \mathbb{R}$ with the property

$$\lim_{t\to\infty} K_t = K_o \iff \lim_{t\to\infty} \int h_k \, d\mu \otimes K_t = \int h_k \, d\mu \otimes K_o \qquad (k \in \mathbb{N})$$

for all sequences $(K_t | t \in \mathbb{N} \cup \{0\})$ from Σ . The proof of the following result on existence of purifying sequences bases on this fact.

<u>1.3.5 Theorem:</u> *Let* K *be a mixed strategy. Then, a sequence* $(\delta_t | t \in \mathbb{N})$ *of discrete strategies with*

(1) $$\lim_{t\to\infty} \delta_t = K$$

(2) $$\operatorname{spt} \delta_t(\omega,.) \subset \operatorname{spt} K(\omega,.) \quad \mu\text{-a.e.} \ (t\in\mathbb{N})$$

exists.

In case μ *is nonatomic, a sequence* $(\varepsilon_{f_t} | t\in\mathbb{N})$ *of pure strategies with*

(1)' $$\lim_{t\to\infty} \varepsilon_{f_t} = K$$

(2)' $$f_t(\omega) \in \operatorname{spt} K(\omega,.) \quad \mu\text{-a.e.} \ (t\in\mathbb{N})$$

exists.

Proof: Let $(h_k | k\in\mathbb{N})$ be given according to the preceding explanations. Then, by 1.3.3 , we find a discrete strategy δ_t for every $t \in \mathbb{N}$ s.t.

$$\int h_k \, d\mu\otimes\delta_t = \int h_k \, d\mu\otimes K \quad (k=1,\ldots,t)$$

as well as $\operatorname{spt} \delta_t(\omega,.) \subset \operatorname{spt} K(\omega,.)$ μ-a.e. is satisfied. From this we conclude that the sequence $(\delta_t | t\in\mathbb{N})$ has properties (1) and (2). The second part of the statement is proved in the same way using 1.3.4 instead of 1.3.3 . □

The denseness result 1.3.5 was proved in a different way by Milgrom and Weber (1985) for the special case of a compact decision space. The advantage of our method is that the approximation of mixed strategies by pure strategies can be brought out in many different ways. For instance, the purifying sequence may be chosen so as to be payoff-equivalent w.r.t. a given function $u \in L^n(\Sigma)$.

1.4 Some topological properties of the set of strategies concentrated on a correspondence

Henceforth, let be given a Polish and locally compact information space (Ω, A, μ), where μ is a probability measure, and let $(S, \mathcal{S})$ be a compact decision space. Further, let d be a distance on S generating the topology of S, and let d^H and d_o^H be the corresponding Hausdorff-distance and Hausdorff-semidistance on $C(S)$, resp.

We first introduce two concepts of measuring distances between correspondences which are based on the Hausdorff-distance and Hausdorff-semidistance, resp.

1.4.1 Definition: *Let* $F, G : \Omega \to C(S)$ *be two* A-$\mathcal{B}(C(S))$*-measurable correspondences. The real numbers*

$$\Delta(F,G) := \int d^H(F(\omega), G(\omega))\, d\mu(\omega)$$

and

$$\Delta_o(F,G) := \int d_o^H(F(\omega), G(\omega))\, d\mu(\omega)$$

are called the expected distance and the expected semidistance of F *and* G*, resp.*

From the definition of the Hausdorff-distance and Hausdorff-semidistance one derives the following facts:

- $\Delta_o(F,G) \leq \Delta(F,G)$
- $\Delta(F,G) = \Delta(G,F)$
- $\Delta(F,G) = 0 \iff F(\omega) = G(\omega)$ μ-a.e.

- $\Delta_o(F,G) = 0 \iff F(\omega) \subset G(\omega)$ μ-a.e.

- $\Delta(F,H) \leq \Delta(F,G) + \Delta(G,H)$

- $\Delta_o(F,H) \leq \Delta_o(F,G) + \Delta_o(G,H)$.

For later use we introduce a concept of distance for r-tuples of pure strategies in a similar way.

<u>1.4.2 Definition:</u> *Let* $(f_1,\ldots,f_r)$ *and* $(g_1,\ldots,g_r)$ *be two r-tuples of pure strategies. Then,*

$$d^*((f_1,\ldots,f_r),(g_1,\ldots,g_r)) = \int \max_{1\leq i\leq r} d(f_i(\omega),g_i(\omega)\, d\mu(\omega)$$

is called the <u>*distance*</u> *of* $(f_1,\ldots f_r)$ *and* $(g_1,\ldots,g_r)$.

We equip the set Σ of all strategies with the weak topology as introduced in 1.1 ; and we define a metric ρ on Σ by

$$\rho(K,L) := \sum_{n\in\mathbb{N}} \frac{1}{2^n} \left| \int h_n \, d\mu\otimes K - \int h_n \, d\mu\otimes L \right| \qquad (K,L \in \Sigma)$$

for some given sequence (h_n) of uniformly continuous functions $h_n : \Omega\times S \to [-1,1]$ w.r.t. a suitable metrization of Ω (cp. A.1.3).

The following lemma shows that ρ is uniformly continuous w.r.t. d^* .

<u>1.4.3 Lemma:</u> *For given* $\varepsilon>0$ *there exists an* $\eta>0$ *s.t.*

$$d^*((f_1,\ldots,f_r),(g_1,\ldots,g_r)) \leq \eta \implies$$

$$\implies \rho(\delta_{(f_1,\ldots,f_r,\alpha_1,\ldots,\alpha_r)},\delta_{(g_1,\ldots,g_r,\alpha_1,\ldots,\alpha_r)}) \leq \varepsilon$$

for arbitrary discrete strategies $\delta_{(f_1,\ldots,f_r,\alpha_1,\ldots,\alpha_r)}$ *and* $\delta_{(g_1,\ldots,g_r,\alpha_1,\ldots,\alpha_r)}$ *and arbitrary* $r \in \mathbb{N}$.

Proof: Let $\varepsilon>0$ be arbitrary. We first record the estimation

$$\text{(i)} \qquad \sum_{n>m} \frac{1}{2^n} \left| \int h_n \, d\mu\otimes K - \int h_n \, d\mu\otimes L \right| \leq \frac{\varepsilon}{3} \qquad (K,L \in \Sigma)$$

for suitable $m \in \mathbb{N}$. Since $h_1,\ldots,h_m$ are uniformly continuous functions, we find an $\eta'>0$ with the property

$$\text{(ii)} \qquad d(s,s') \leq \eta' \implies \max_{1\leq n\leq m} |h_n(\omega,s) - h_n(\omega,s')| \leq \frac{\varepsilon}{3} \qquad (\omega\in\Omega,\ s,s' \in S).$$

Now, let be given two discrete strategies

$\delta_1 := \delta_{(f_1,\ldots,f_r,\alpha_1,\ldots,\alpha_r)}$ and $\delta_2 := \delta_{(g_1,\ldots,g_r,\alpha_1,\ldots,\alpha_r)}$.

Observing (i), we obtain the estimation

$$\rho(\delta_1,\delta_2) = \sum_{n\in\mathbb{N}} \frac{1}{2^n} \left| \int \sum_{i=1}^{r} \alpha_i(\omega)\, h_n(\omega,f_i(\omega))\, d\mu - \right.$$

$$\left. - \int \sum_{i=1}^{r} \alpha_i(\omega)\, h_n(\omega,g_i(\omega))\, d\mu \right| \leq$$

$$\leq \left(\sum_{n=1}^{m} \frac{1}{2^n} \int \sum_{i=1}^{r} \alpha_i \,|h_n(.,f_i(.)) - h_n(.,g_i(.))|\, d\mu\right) + \frac{\varepsilon}{3}.$$

Setting $z_i(.) := \max_{1\leq n\leq m} |h_n(.,f_i(.)) - h_n(.,g_i(.))| \quad (i=1,\ldots,r)$

we obtain $|z_i| \leq 2$ and the following chain of inequalities

$$\rho(\delta_1,\delta_2) \leq \int \sum_{i=1}^{r} \alpha_i z_i \, d\mu + \frac{\varepsilon}{3} \leq$$

$$\leq \int_{\{\max_{1\leq i\leq r} d(f_i,g_i)\leq\eta'\}} \sum \alpha_i z_i \, d\mu +$$

$$+ \int_{\{\max_{1\leq i\leq r} d(f_i,g_i)>\eta'\}} \sum \alpha_i z_i \, d\mu + \frac{\varepsilon}{3} \leq$$

$$\underset{(ii)}{\leq} \frac{2}{3}\varepsilon + 2\mu\{\max_{1\leq i\leq r} d(f_i,g_i) > \eta'\} \leq$$

$$\leq \frac{2}{3}\varepsilon + \frac{2}{\eta'}\int \max_{1\leq i\leq r} d(f_i,g_i)\, d\mu =$$

$$= \frac{2}{3}\varepsilon + \frac{2}{\eta'} d^*((f_1,\dots,f_r),(g_1,\dots,g_r)) .$$

Choosing $\eta := \frac{1}{6}\eta'\varepsilon$, we get the desired result, since η' does not depend on δ_1 and δ_2. □

In the sequel, we investigate the set of strategies which are concentrated on a correspondence. We first derive a useful continuity result for the expected payoff w.r.t. a certain type of payoff-function.

1.4.4 Definition: *Every function* $u : \Omega\times S \to \mathbb{R}^n$ *with the properties*

(1) $u(.,s)$ *is* $\mathcal{A}$*-measurable* $(s\in S)$,

(2) $u(\omega,.)$ *is continuous* $(\omega\in\Omega)$,

(3) $|u_i(\omega,s)| \leq g(\omega)$ $(\omega\in\Omega,\ s\in S;\ i=1,\dots,n)$,

is called a *μ-Carathéodory-function (abbreviated:* *μ-C-function).*

Every μ-C-function is $\mathcal{A}\otimes\mathcal{S}$-measurable (cp. A.1.8) and is therefore $\mu\otimes K$-integrable for arbitrary strategies K. Moreover, this type of payoff-function yields continuous expected payoffs.

1.4.5 Theorem: *Let be given a μ-C-function* $u : \Omega\times S \to \mathbb{R}^n$.
Then, the function U *defined by*

$$U(K) = \int u \, d\mu\otimes K$$

is continuous on the set Σ *of all strategies* K.

Proof: Let $\varepsilon > 0$ be given. First, there exists a μ-integrable function $g : \Omega \to \mathbb{R}$ with

$$|u_i(.,s)| \leq g \quad (s \in S;\ i=1,\ldots,n).$$

Since Ω is a Polish space, a compact set $C \subset \Omega$ can be found satisfying

$$\int_{\Omega - C} g \, d\mu \leq \varepsilon .$$

Moreover, by A.1.9, C can be chosen in such a way that u is continuous on $C \times S$ and g is continuous on C. The function $(u_i + g) 1_{C \times S}$ is therefore un upper semicontinuous function for $i=1,\ldots,n$. Hence, the function $\overline{V}_i : \Sigma \to \mathbb{R}$ defined by

$$\overline{V}_i(K) := \int_{C \times S} (u_i + g) \, d\mu \otimes K \quad (K \in \Sigma)$$

is upper semicontinuous on Σ. A similar argumentation shows that the function $\underline{V}_i : \Sigma \to \mathbb{R}$ defined by

$$\underline{V}_i(K) := \int_{C \times S} (u_i - g) \, d\mu \otimes K \quad (K \in \Sigma)$$

is lower semicontinuous on Σ. As the equality

$$\overline{V}_i = \underline{V}_i + 2 \int_C g \, d\mu$$

holds, $\overline{V}_i$ and $\underline{V}_i$ are upper and lower continuous functions, and are therefore continuous. Define $V_i : \Sigma \to \mathbb{R}$ by

$$V_i(K) := \int_{C \times S} u_i \, d\mu \otimes K \quad (K \in \Sigma) .$$

Then, since $V_i = \frac{1}{2}(\overline{V}_i + \underline{V}_i)$ holds, V_i is a continuous function on Σ and satisfies

$$|V_i(K) - U_i(K)| = \left|\int_{(\Omega-C)\times S} u_i \, d\mu\otimes K\right| \leq$$

$$\leq \int_{\Omega-C} g \, d\mu \leq \varepsilon \qquad (K\in\Sigma)$$

This shows that U_i can be uniformly approximated by continuous functions and is therefore itself a continuous function. □

The above theorem will be used to prove the following fact.

<u>1.4.6 Lemma:</u> *Let* $F : \Omega \to C(S)$ *be an* A-$B(C(S))$*-measurable correspondence. Then, the set*

$$\Sigma_F := \{K\in\Sigma \mid K(\omega, F(\omega)) = 1 \quad \mu\text{-a.e.}\}$$

is a nonempty convex and compact subset of Σ .

Proof: The existence of $K \in \Sigma_F$ follows from A.2.9 and A.2.10 . The convexity of Σ_F is obvious. To prove compactness of Σ_F , we first note that

$$\Sigma_F = \{K\in\Sigma \mid \int d(s,F(\omega)) \, d\mu\otimes K = 0\} .$$

Since $(\omega,s) \mapsto d(s,F(\omega))$ is a μ-C-function on $\Omega\times S$ (cp. also A.2.5), this representation of Σ_F shows together with Theorem 1.4.5 that Σ_F is a closed subset of Σ. It suffices now to verify that the set $\{\mu\otimes K \mid K\in\Sigma\}$ is relatively compact in the set of all probability measures on $A\otimes S$. To this end, fix a compact set $C \subset \Omega$ with $\mu(\Omega-C) \leq \varepsilon$, which exists, since Ω is Polish. Then

$$\mu\otimes K(C\times S) \geq 1-\varepsilon \qquad (K\in\Sigma) .$$

By the Theorem of Prohorov (A.1.5) this proves that the set $\{\mu\otimes K \mid K\in\Sigma\}$ is a relatively compact subset of the set of all probability measures on $A\otimes S$. □

Of course, Lemma 1.4.6 implies also compactness of Σ: choose especially $F \equiv S$ in 1.4.6 .

We consider now the Hausdorff-distance ρ^H (deduced from ρ) on $C(\Sigma)$ and the corresponding Hausdorff-semidistance ρ_o^H on $C(\Sigma)$. The following theorem points out that there is a close connection between the expected semidistance of correspondences and the semidistance on $C(\Sigma)$.

<u>1.4.7 Theorem:</u> *For arbitrarily chosen* $\varepsilon>0$ *an* $\eta>0$ *exists such that*

$$\Delta_o(F,G) \leqq \eta \implies \rho_o^H(\Sigma_F,\Sigma_G) \leqq \varepsilon$$

holds for every two measurable correspondences $F,G:\Omega \to C(S)$ *at a time.*

Proof: Let $\varepsilon>0$ be given. Then, by Lemma 1.4.3 , there exists an $\eta>0$ with the property

$$\text{(i)} \quad \begin{aligned} & d^*((f_1,\dots,f_r),(g_1,\dots,g_r)) \leqq \eta \implies \\ & \implies \rho(\delta_{(f_1,\dots,f_r,\alpha_1,\dots,\alpha_r)},\delta_{(g_1,\dots,g_r,\alpha_1,\dots,\alpha_r)}) \leqq \frac{\varepsilon}{2} \end{aligned}$$

for arbitrary discrete strategies $\delta_{(f_1,\dots,f_r,\alpha_1,\dots,\alpha_r)}$ and $\delta_{(g_1,\dots,g_r,\alpha_1,\dots,\alpha_r)}$.

Let now $F,G:\Omega \to C(S)$ be two measurable correspondences; and choose $K\in\Sigma_F$. By Theorem 1.3.5, a discrete strategy

$\delta_{(f_1,\dots,f_r,\alpha_1,\dots,\alpha_r)}$ exists s.t.

(ii) $\qquad \rho(\delta_{(f_1,\dots,f_r,\alpha_1,\dots,\alpha_r)},K) \leq \frac{\varepsilon}{2}$,

(iii) $\qquad f_i(\omega) \in \operatorname{spt} K(\omega,.) \quad \mu\text{-a.e.} \quad (i=1,\dots,r)$.

The latter relation implies

(iv) $\qquad f_i(\omega) \in F(\omega) \quad \mu\text{-a.e.} \quad (i=1,\dots r)$.

For $i=1,\dots,r$ we set

$$G_i^*(\omega) := \{s \in G(\omega) \mid d(f_i(\omega),s) = d(f_i(\omega),G(\omega))\} .$$

Then, the correspondence $G_i^* : \Omega \to C(S)$ has a measurable graph by A.2.5 and A.2.9 , and admits therefore a measurable selection $g_i : \Omega \to S$, i.e. $g_i(\omega) \in G_i^*(\omega)$ holds μ-a.e. By definition of g_i $(i=1,\dots,r)$, the discrete strategy $\delta_{(g_1,\dots,g_r,\alpha_1,\dots,\alpha_r)}$ satisfies

$$d^*((f_1,\dots,f_r),(g_1,\dots,g_r)) = \int \max_{1\leq i\leq r} d(f_i(.),g_i(.))\, d\mu =$$

$$= \int \max_{1\leq i\leq r} d(f_i(.),G(.))\, d\mu \leq$$

$$\leq \int \max_{s\in F(\omega)} d(s,G(\omega))\, d\mu = \Delta_o(F,G) .$$

In case $\Delta_o(F,G) \leq \eta$, we conclude from this estimation together with (i) and (ii) that

$$\rho(K,\delta_{(g_1,\dots,g_r,\alpha_1,\dots,\alpha_r)}) \leq \varepsilon$$

holds. Since $K \in \Sigma_F$ was arbitrarily chosen, and since $\delta_{(g_1,\dots,g_r,\alpha_1,\dots,\alpha_r)} \in \Sigma_G$, this proves

$$\rho_o(\Sigma_F,\Sigma_G) = \max_{K\in\Sigma_F} \rho(K,\Sigma_G) \leq \varepsilon . \qquad \square$$

The following corollary is an immediate consequence of the relation

$$\Delta_o(F,G) \leqq \Delta(F,G)$$

and the definition of the Hausdorff-distance.

<u>1.4.8 Corollary:</u> *For arbitrarily chosen* $\varepsilon>0$ *an* $\eta>0$ *exists such that*

$$\Delta(F,G) \leqq \eta \implies \rho^H(\Sigma_F,\Sigma_G) \leqq \varepsilon$$

is satisfied for every two measurable correspondences $F,G : \Omega \to C(S)$ *at a time.*

With these preparations we are now in a position to prove a result which will be utilized in connection with the proof of existence of an equilibrium for the game described in 1.1.

<u>1.4.9 Theorem:</u> *In addition to the assumptions made in the beginning of this section, let be given a metric space* E. *Let*

$$F : \Omega\times E \to C(S)$$

be a correspondence with the properties

- $F(\omega,.)$ *is continuous at* $t_o \in E \quad (\omega\in\Omega)$,
- $F(.,t)$ *is* $\mathcal{A}$-$\mathcal{B}(C(S))$-*measurable* $\quad (t\in E)$.

Then, the correspondence $\Pi : E \to C(\Sigma)$ *defined by*

$$\Pi(t) := \Sigma_{F(.,t)} \qquad (t\in E)$$

is continuous at t_o *(w.r.t.* ρ^H *on* $C(\Sigma)$*).*

Proof: Let (t_n) be a sequence from E converging to t_o. By the dominated convergence theorem, the continuity of the correspondences $F(\omega,.)$ $(\omega\in\Omega)$ at t_o implies

$$\lim_{n\to\infty} \Delta(F(.,t_o),F(.,t_n)) =$$

$$= \lim_{n\to\infty} \int d^H(F(.,t_o),F(.,t_n))\, d\mu =$$

$$= \int \lim_{n\to\infty} d^H(F(.,t_o),F(.,t_n))\, d\mu = 0 .$$

Together with Corollary 1.4.8, this proves

$$\lim_{n\to\infty} \rho^H(\Sigma_{F(.,t_o)},\Sigma_{F(.,t_n)}) = 0 . \qquad \square$$

Finally, we give a useful characterization of strategies maximizing the expected payoff within the set of strategies which are concentrated on a correspondence. Roughly spoken, the result shows that local and global maximization are equivalent.

<u>1.4.10 Lemma:</u> *Let* $u : \Omega\times S \to \mathbb{R}$ *be a* μ*-C-function; let be given an* A-$B(C(S))$*-measurable correspondence* $F : \Omega \to C(S)$*, and let* $K \in \Sigma_F$ *be arbitrary. Then, the following properties are equivalent:*

(1) $\int u \, d\mu\otimes K = \max_{L\in\Sigma_F} \int u \, d\mu\otimes L$;

(2) *for* μ*-almost every* $\omega\in\Omega$ *the equation*

$$u(\omega,s) = \max_{t\in F(\omega)} u(\omega,t)$$

is satisfied $K(\omega,.)$ -a.e.

Proof: For $L \in \Sigma_F$, we always have

(i) $u(\omega,s) \leq \max_{t\in F(\omega)} u(\omega,t)$ $L(\omega,.)$-a.e. μ-a.e.

Integration of this inequality w.r.t. $\mu\otimes L$ on both sides, and integration of the equality in (2) w.r.t. $\mu\otimes K$ on both sides shows that conclusion (2) $\Longrightarrow$ (1) holds.

Now, to prove the converse conclusion , let (1) be satisfied. Define the correspondence $G : \Omega \to C(S)$ by

$$G(\omega) := \{s\in F(\omega) \,|\, u(\omega,s) = v(\omega)\} \quad (\omega\in\Omega),$$

where $v : \Omega \to \mathbb{R}$ is an A-measurable function with the property

$$v(\omega) = \max_{t\in F(\omega)} u(\omega,t) \quad \mu\text{-a.e.}$$

(according to A.1.10, such a function exists). Then, G has a measurable graph and allows therefore a measurable selection $f : \Omega \to S$. The pure strategy ε_f satisfies

(ii) $$\int u \, d\mu\otimes\varepsilon_f = \int \max_{t\in F(\omega)} u(\omega,t) \, d\mu .$$

On the other hand, since $K \in \Sigma_F$, we have

(iii) $$\int u \, d\mu\otimes K = \int \Big(\int_{F(\omega)} u(\omega,s) \, K(\omega,ds)\Big) \, d\mu;$$

and, since (1) is satisfied, we have

$$\int u \, d\mu\otimes K \geq \int u \, d\mu\otimes\varepsilon_f .$$

Together with (ii) and (iii), this inequality proves

$$u(\omega,s) = \max_{t\in F(\omega)} u(\omega,t) \quad K(\omega,.)\text{-a.e.} \quad \mu\text{-a.e.}$$

Hence, (2) is satisfied. □

1.5 Theorems on existence of an equilibrium

In this section, we first prove a general result on existence of an equilibrium for the game described in 1.1 . Thereby, we assume that the preferences of players are given by preference representation functions. A further existence theorem concerns the case, when preferences are laid down by payoff-functions.

At the same time, an essential aspect of games with incomplete information will be illustrated, namely, that is the advantage for players to use pure strategies instead of mixed strategies without crucial effects on the preferences.

Throughout this section, we assume that

(1) $(\Omega_i, \mathcal{A}_i, \mu_i)$ is a Polish and locally compact information space;

(2) μ_i is a probability measure;

(3) $(S_i, \mathcal{S}_i)$ is a compact decision space w.r.t. some metric d_i;

(4) R_i is a Carathéodory constraint correspondence with values in $\mathcal{C}(S_i)$, i.e.,

- $R_i(.,K)$ is an $\mathcal{A}_i$-$\mathcal{B}(\mathcal{C}(S_i))$-measurable correspondence for every $K \in \bigtimes_{j=1}^{N} \Sigma_j$,
- $R_i(\omega_i,.)$ is a continuous correspondence for every $\omega_i \in \Omega_i$,

for every player i. Further, let v_i be a preference representation function for player i.

With these assumptions we obtain a first result on existence of an equilibrium.

<u>1.5.1 Theorem:</u> *If assumptions (1)-(4) are satisfied, then the game has an equilibrium point.*

Proof: Lemma 1.4.6 shows that the space $\Sigma := \bigtimes_{i=1}^{N} \Sigma_i$ is nonempty, convex and compact. Further, setting

$$\Pi_i(K) := \{L_i \in \Sigma_i | L_i \text{ is admissible w.r.t. } K\} =$$
$$= \{L_i \in \Sigma_i | L_i(\omega_i, R_i(\omega_i, K)) = 1 \quad \mu_i\text{-a.e.}\}$$

for $K \in \Sigma$, as a consequence of 1.4.6 and 1.4.9 , we obtain a continuous, convex- and compact-valued correspondence $\Pi_i : \Sigma \to C(\Sigma_i)$. Let the correspondence $\Pi^* : \Sigma \to C(\Sigma)$ be given by

$$\Pi^*(K) := \bigtimes_{i=1}^{N} \{L_i \in \Pi_i(K) | v_i(K, L_i) = \max_{L_i' \in \Pi_i(K)} v_i(K, L_i')\} \qquad (K \in \Sigma).$$

Then, because v_i is continuous, A.2.4 shows that Π^* is a (nonempty) compact-valued, upper semicontinuous correspondence. Moreover, since v_i is quasiconcave in the second variable, Π^* is convex-valued. Therefore, the Fixed Point Theorem A.5.1 applies to Π^* and delivers a strategy $K \in \Sigma$ with

$$K \in \Pi^*(K) .$$

By definition of Π_i and Π^*, K_i is admissible w.r.t. K for every player i and satisfies

$$v_i(K, L_i) \leq v_i(K, K_i) = 0$$

for abitrary admissible strategies L_i w.r.t. K . Hence, K is an equilibrium point. □

We are going now to discuss the question of existence of approximate equilibrium in pure strategies. For that purpose we remember the concept of expected semidistance of two correspondences $F_i, G_i : \Omega_i \to S_i$ as introduced in 1.4.1 and denote it by Δ_o^i $(i \in I)$, i.e.

$$\Delta_o^i(F_i,G_i) = \int d_{io}^H (F_i(.),G_i(.))\, d\mu_i ,$$

when d_{io}^H is the Hausdorff-semidistance on $C(S_i)$.

Apart from this measure for the deviation of two correspondences we will consider the μ_i-essential maximal semidistance defined by

$$\Delta_{ess}^i (F_i,G_i) =$$

$$= \inf\{\alpha \in \mathbb{R} \mid d_{io}^H (F_i(.),G_i(.)) \leq \alpha \quad \mu_i\text{-a.e.}\} .$$

Basing on these two deviation measures we introduce a modified equilibrium concept.

1.5.2 Definition: *Let be given* $K = (K_1,\dots,K_N) \in \Sigma$. *For* $i \in I$ *let* Δ^i *be one of functions* Δ_o^i *and* Δ_{ess}^i *in each case.*

(1) *Let* $\delta \geq 0$ *be arbitrary. A strategy* L_i *of player* i *is called a* δ*-admissible strategy w.r.t.* K, *iff*

$$\Delta^i(\text{spt } L_i, R_i(.,K)) \leq \delta$$

holds for the correspondence $\text{spt } L_i : \omega_i \to \text{spt } L_i(\omega_i,.)$.

(2) *Let* $\varepsilon,\delta \geq 0$ *be arbitrary. The strategy N-tuple* K *is called an* $\varepsilon\delta$*-equilibrium point of the game, if* K_i *is a* δ*-admissible strategy w.r.t.* K *of every player* i, *and*

$$v_i(K,L_i) \leq \varepsilon$$

holds for all δ-admissible strategies L_i *w.r.t.* K .

The following theorem shows that a pure strategy equilibrium point in the above sense exists.

1.5.3 Theorem: *Let assumptions (1)-(4) be satisfied; and let* μ_i *be a nonatomic measure for every player i. Moreover, we assume that* Δ_o^i *is the underlying deviation measure. Then, for every* $\varepsilon>0$ *there is some* $\delta_o>0$ *such that for every* $\delta>0$ *with* $\delta\leq\delta_o$ *there exists an* εδ*-equilibrium point in pure strategies.*

Proof: We equip each set Σ_i with a metric ρ_i of the type as defined in 1.4 and denote the corresponding Hausdorff-semi-distance on $C(\Sigma_i)$ by ρ_{io}^H. Moreover, according to the notation introduced in 1.4.6 , we set

$$\Sigma_{i,F_i} := \{K_i\in\Sigma_i | K_i(\omega_i,F_i(\omega_i)) = 1 \quad \mu_i\text{-a.e.}\}$$

for a measurable correspondence $F_i : \Omega_i \to C(S_i)$. By Theorem 1.5.1 an equilibrium point $\overline{K} \in \Sigma$ exists. Hence, we have

$$v_i(\overline{K},L_i) \leq 0 \quad \text{for every } L_i \in \Sigma_{i,R_i(\cdot,\overline{K})} \quad (i\in I) .$$

Since $\Sigma \times \Sigma_i$ is compact metric, the function v_i is uniformly continuous on $\Sigma \times \Sigma_i$ (for a suitable metrization). Therefore, a $\delta'>0$ exist such that

$$\rho_i(K_i,\overline{K}_i) \leq \delta', \quad \rho_i(L_i,\Sigma_{i,R_i(\cdot,\overline{K})}) \leq \delta' \quad (i\in I) \quad \Longrightarrow$$

$$\text{(i)} \qquad \Longrightarrow \quad v_i(K,L_i) \leq \varepsilon \quad (i\in I)$$

$$(K = (K_1,\ldots,K_N) \in \Sigma, \; L = (L_1,\ldots,L_N) \in \Sigma) .$$

Theorem 1.4.7 shows that there is some $\delta_o>0$ with the property

$$\text{(ii)} \qquad \Delta_o^i(F_i,G_i) \leq 2\delta_o \quad \Longrightarrow \quad \rho_{io}^H(\Sigma_{i,F_i},\Sigma_{i,G_i}) \leq \delta'$$

for every two A_i-$\mathcal{B}(\mathcal{C}(S_i))$-measurable correspondences $F_i, G_i : \Omega_i \to \mathcal{C}(S_i)$ $(i \in I)$. Fix now $\delta > 0$ with $\delta \leq \delta_o$. Combining Theorem 1.3.5 with assumption (4) and applying the dominated convergence theorem, we obtain a strategy N-tuple $K^* = (\varepsilon_{f_1}, \ldots, \varepsilon_{f_N})$ with the properties

(iii) $\Delta_o^i(R_i(.,K^*), R_i(.,\overline{K})) \leq \delta$, $\Delta_o^i(R_i(.,\overline{K}), R_i(.,K^*)) \leq \delta$

(iv) $f_i(\omega_i) \in \operatorname{spt} \overline{K}_i(\omega_i,.)$ μ_i-a.e.

and

(v) $\rho_i(\varepsilon_{f_i}, \overline{K}_i) \leq \delta'$

for every $i \in I$. Since $\overline{K}$ is an equilibrium point, we conclude from (iv)

(vi) $f_i(\omega_i) \in \operatorname{spt} \overline{K}_i(\omega_i,.) \subset R_i(\omega_i, \overline{K})$ μ_i-a.e.

The triangle inequality for Δ_o^i together with (iii) and (vi) shows that

(vii) $\Delta_o^i(\{f_i(.)\}, R_i(.,K^*)) \leq \Delta_o^i(R_i(.,\overline{K}), R_i(.,K^*)) \leq \delta$.

Now, let L_i be a δ-admissible strategy of player i w.r.t. K^*, i.e.

(viii) $\Delta_o^i(\operatorname{spt} L_i, R_i(.,K^*)) \leq \delta$.

A further application of the triangle inequality for Δ_o^i shows together with (iii) that

$$\Delta_o^i(\operatorname{spt} L_i, R_i(.,\overline{K})) \leq$$
$$\leq \Delta_o^i(\operatorname{spt} L_i, R_i(.,K^*)) + \Delta_o^i(R_i(.,K^*), R_i(.,\overline{K})) \leq 2\delta$$

holds. Since $\delta \leq \delta_o$, we derive from (viii) and (ii)

$$\rho_{io}^H(\Sigma_{i,\operatorname{spt} L_i}, \Sigma_{i,R_i(.,\overline{K})}) \leq \delta'.$$

Obviously, we have $L_i \in \Sigma_{i,\mathrm{spt}\, L_i}$, and therefore

$$\rho_i(L_i, \Sigma_{i,R_i(\cdot,\overline{K})}) \leq \delta' .$$

Together with (v) and (i) we conclude from this that

$$v_i(K^*, L_i) \leq \varepsilon$$

is satisfied. Observing (vii), we have proved that K^* is an $\varepsilon\delta$-equilibrium point. □

In order to derive a similar result for the deviation measure Δ^i_{ess} instead of Δ^i_o, we first strengthen assumption (4) in the following sense:

(4)' R_i is a uniform Carathéodory constraint correspondence, i.e., R_i is a Carathéodory constraint correspondence and

- $R_i(\omega_i,\cdot)$ is a uniformly continuous correspondence w.r.t. d_i^H on $C(S_i)$ for every $\omega_i \in \Omega_i$.

<u>1.5.4 Theorem:</u> *Let assumptions (1)-(3) and (4)' be satisfied; and let μ_i be a nonatomic measure for every player i. Then, Theorem 1.5.3 holds with Δ^i_o replaced by Δ^i_{ess}, accordingly.*

Sketch of proof: The proof of Theorem 1.5.3 can be adopted apart from few modifications. Instead (iii) and (vii) we obtain the stronger estimations

(iii)' $$d_i^H(R_i(\omega_i,K^*), R_i(\omega_i,\overline{K})) \leq \delta \quad (\omega_i \in \Omega_i)$$

and

(vii)' $$d_i(f_i(\omega_i), R_i(\omega_i,K^*)) \leq \delta \quad \mu_i\text{-a.e.}$$

Since every δ-admissible strategy L_i of player i w.r.t. K^* satisfies

$$d_{io}^{H}(\text{spt } L_i, R_i(.,K^*)) \leq \delta \quad \mu_i\text{-a.e.},$$

we have

$$\Delta_o^i(\text{spt } L_i, R_i(.,K^*)) \leq \delta$$

which is statement (viii) in the proof of Theorem 1.5.3. The remaining part of the proof is now completely the same as in 1.5.3. □

In many situations preferences of players are derived from payoff-functions

$$u_i : \bigtimes_{j=1}^{N} \Omega_j \times \bigtimes_{j=1}^{N} S_j \to \mathbb{R} \quad (i \in I).$$

Since player i does not know the informations of his counter players, the game can also be considered as a game with unknown payoff-functions. Let be given an information distribution μ on $A = \bigtimes_{i=1}^{N} A_i$, which we assume to be known to all players. We further assume that μ is a finite measure with a density function $f : \bigtimes \Omega_j \times \bigtimes S_j \to \mathbb{R}_+$ w.r.t. the product measure $\mu_1 \otimes \ldots \otimes \mu_N$, i.e.

$$\mu = f(\mu_1 \otimes \ldots \otimes \mu_N). \tag{5}$$

Moreover, we claim that

$$u_i \text{ is a } \mu\text{-C-function} \tag{6}$$

for every player i (cp. 1.4.4).

For every strategy N-tuple $(K_1, \ldots, K_N)$ let the stochastic kernel $K_1 \otimes \ldots \otimes K_N$ be defined by

$$K_1 \otimes \ldots \otimes K_N(\omega_1,\ldots,\omega_N,A_1\times\ldots\times A_N) := K_1(\omega_1,A_1)\cdot\ldots\cdot K_N(\omega_N,A_N)$$

$$((\omega_1,\ldots,\omega_N)\in\Omega,\ A_i\in\mathcal{A}_i\ (i\in I)) .$$

The function $U_i : \Sigma \to \mathbb{R}$ defined by

$$U_i(K_1,\ldots,K_N) := \int u_i \, d\mu\otimes(K_1\otimes\ldots\otimes K_N) \quad (K_j\in\Sigma_j\ (j\in I))$$

is called <u>expected payoff-function</u> of player i . We first realize the following facts.

<u>1.5.5 Lemma:</u> *The function* $U_i : \Sigma \to \mathbb{R}$ *is continuous for player* i *.*

Proof: From assumption (5) we derive

$$U_i(K_1,\ldots,K_N) = \int u_i(\omega,s)\, f(\omega)\, d(\mu_1\otimes\ldots\otimes\mu_N)\otimes(K_1\otimes\ldots\otimes K_N) =$$

$$= \int u_i(\omega,s)\, f(\omega)\, d \bigotimes_{j=1}^{N} (\mu_j\otimes K_j)$$

$$((K_1,\ldots,K_N) \in \Sigma) .$$

As a consequence of (5) and (6), the function u_i^* defined by

$$u_i^*(\omega,s) := u_i(\omega,s)\, f(\omega) \quad (\omega\in \times\,\Omega_j,\ s\in \times\,S_j)$$

is a $\bigotimes_{j=1}^{N} \mu_j$-C-function. Since the mapping

$$(\mu_1\otimes K_1,\ldots,\mu_N\otimes K_N) \mapsto \bigotimes_{j=1}^{N} \mu_j\otimes K_j$$

is continuous w.r.t. the corresponding weak topologies (cp. A.1.6), continuity of U_i follows from (i) and 1.4.5 . □

We set now

$$v_i(K,L_i) := U_i(K_1,\ldots,K_{i-1},L_i,K_{i+1},\ldots,K_N) - U_i(K_1,\ldots,K_N)$$

$$((K_1,\ldots,K_N)\in\Sigma,\ L_i\in\Sigma_i,\ i\in I) .$$

By this means, $v_i : \Sigma \times \Sigma_i \to \mathbb{R}$ becomes a continuous function (cp. 1.5.5). Moreover, we have

$$v_i(K,K_i) = 0 \quad (K \in \Sigma) ,$$

and $v_i(K,.)$ is a quasiconcave function. Hence, v_i is a PRF for player i. This gives rise to an application of the Existence Theorem 1.5.1 to the present situation. Nevertheless, we have first to discuss, in how far the equilibrium concepts established here are compatible with the usual equilibrium concept in noncooperative game theory.

<u>1.5.6 Definition:</u> *Let Γ be the noncooperative game given by $\Sigma_1,\ldots,\Sigma_N$, $U_1,\ldots,U_N$ and $R_1,\ldots,R_N$.*

(1) *A strategy N-tuple $K := (K_1,\ldots,K_N)$ is called an <u>equilibrium point of Γ</u>, if K_i is admissible w.r.t. K, and*

$$U_i(K_1,\ldots,K_{i-1},L_i,K_{i+1},\ldots,K_N) \le U_i(K_1,\ldots,K_N)$$

holds for every admissible strategy L_i w.r.t. K and every player i.

(2) *Let Δ^i be one of the deviation measures Δ^i_o and Δ^i_{ess}. For given $\varepsilon,\delta \ge 0$, a strategy N-tuple $K := (K_1,\ldots,K_N)$ is called an <u>$\varepsilon\delta$-equilibrium point of Γ</u>, iff K_i is δ-admissible w.r.t. K, and*

$$U_i(K_1,\ldots,K_{i-1},L_i,K_{i+1},\ldots,K_N) \le U_i(K_1,\ldots,K_N) + \varepsilon$$

holds for every δ-admissible strategy L_i w.r.t. K and for every player i.

Obviously, the equilibrium concepts introduced in 1.1.2 and 1.5.2, resp., coincide with those given in 1.5.6, when preferences of players are represented by the PRFs v_i as defined above. Therefore, the Existence Theorems 1.5.1, 1.5.3 and 1.5.4 may be formulated in terms of the game Γ as given in Definition 1.5.6. In doing so we obtain a modified version of a result proved by Milgrom and Weber (1985). The result to be presented now, is in so far an extension of the result of Milgrom and Weber as there are constraint correspondences affecting the game.

1.5.7 Theorem: *If assumptions (1)-(6) are satisfied, then the game Γ as given in 1.5.6 has an equilibrium point. Moreover, if μ_i is nonatomic* ($i \in I$), *then for every $\varepsilon>0$ there is some $\delta_o>0$ such that an $\varepsilon\delta$-equilibrium point in pure strategies w.r.t. Δ_o^i* ($i \in I$) *exists for every $\delta>0$ with $\delta \leq \delta_o$. If (4)' is satisfied, the same holds true for Δ_{ess}^i instead of Δ_o^i* ($i \in I$).

We finally mention a useful characterization of equilibrium points. To this end we introduce the conditional payoff $w_i : \Omega_i \times S_i \times \mathop{\times}_{j \neq i} \Sigma_j \to \mathbb{R}$ of player i by

$$w_i(\omega_i, s_i, K_{-i}) := \int u_i(\omega, s)\, f(\omega)\, d \bigotimes_{j \neq i} \mu_j \otimes K_j$$

$$(\omega = (\omega_1, \ldots, \omega_N) \in \mathop{\times} \Omega_j,$$

$$s = (s_1, \ldots, s_N) \in \mathop{\times} S_j,$$

$$K_{-i} = (K_1, \ldots, K_{i-1}, K_{i+1}, \ldots, K_N) \in \mathop{\times}_{j \neq i} \Sigma_j).$$

Thus, we get

$$U_i(K_1,\dots,K_N) = \int w_i(\omega_i, s_i, K_{-i})\, d\mu \otimes K_i \,.$$

As a consequence of (5) and (6), the conditional payoff $w_i(.,.,K_{-i})$ is a μ_i-C-function. As an application of 1.4.10 we obtain a following result.

<u>1.5.8 Lemma:</u> *A strategy N-tuple* $K=(K_1,\dots,K_N)$ *is an equilibrium point, if and only if* K_i *is K-admissible for every player i and*

$$w_i(\omega_i, s_i, K_{-i}) = \max_{s_i' \in R_i(\omega_i, K)} w_i(\omega_i, s_i', K_{-i}) \qquad K_i(\omega_i,.)\text{-a.e.}$$

holds for μ_i*-a.e.* $\omega_i \in \Omega_i$.

1.6 On existence of pure strategy equilibrium

While we have been concerned with some results on existence of approximate equilibrium in pure strategies in the last section, we are going now to consider games which allow exact equilibria in pure strategies. Unfortunately, even in case of a nonatomic information distribution, such an equilibrium may not exist. This will be demonstrated by an example in Section 3.1 . However, for certain types of payoff-functions and constraints the purification problem proves to have a satisfactory solution.

Like in the last section we assume that

(1) (Ω_i, A_i, μ_i) is a Polish and locally compact information space;

(2) μ_i is a nonatomic probability measure;

(3) $(S_i, \mathcal{S}_i)$ is a compact decision space for every player i.

We define the information distribution by $\bigotimes_{i=1}^{N} \mu_i$, i.e., the informations of players are independently distributed. Especially, assumption (5) in section 1.5 is satisfied in the given situation.

1.6.1 Definition: *Let* $v_i : \Omega_i \times S_i \to \mathbb{R}$ $(i \in I)$ *be* μ_i*-C-functions. Then, the function* $v_1 \otimes \ldots \otimes v_N : \times \Omega_i \times \times S_i \to \mathbb{R}$ *defined by*

$$v_1 \otimes \ldots \otimes v_N(\omega_1, \ldots, \omega_N, s_1, \ldots, s_N) := \prod_{i=1}^{N} v_i(\omega_i, s_i)$$

$$(\omega_i \in \Omega_i, s_i \in S_i;\ i \in I)$$

is called the <u>*tensor product of the functions*</u> $v_1,\dots,v_N$.

Every function $v: \mathsf{X}\Omega_i \times \mathsf{X} S_i \to \mathbb{R}$ *of the type*

$$v = \sum_{k=1}^{r} v_1^{(k)} \otimes \dots \otimes v_N^{(k)}$$

with tensor products $v_1^{(k)} \otimes \dots \otimes v_N^{(k)}$ $(k=1,\dots,r)$ *is called a* <u>*sum of tensor products*</u>.

A short calculation shows that every sum of tensor products is a $\bigotimes_{i=1}^{N} \mu_i$-C-function.

Next, we introduce a special type of constraint correspondences.

<u>1.6.2 Definition:</u> *For some* $n \in \mathbb{N}$ *let*

$$R_i^*: \Omega_i \times \mathbb{R}^n \to C(S_i)$$

be a correspondence for which

- $R_i^*(.,x)$ *is a measurable correspondence for every* $x \in \mathbb{R}^n$

and

- $R_i^*(\omega_i,.)$ *is a continuous correspondence for every* $\omega_i \in \Omega_i$.

Then, we call R_i^* *a* <u>*standard correspondence*</u>.

Suppose, for every $i \in I$ *there is an* $n_i \in \mathbb{N}$, and there are μ_i*-C-functions* $w_i: \Omega_i \times S_i \to \mathbb{R}^{n_i}$ *and a suitable standard correspondence* R_i^* *such that*

$$R_i(\omega_i,K_1,\dots,K_N) = R_i^*(\omega_i,\int w_1\, d\mu_1 \otimes K_1,\dots,\int w_N\, d\mu_N \otimes K_N)$$

$$(\omega_i \in \Omega_i,\ (K_1,\dots,K_N) \in \Sigma),$$

then we say that the family $(R_i | i \in I)$ *is* <u>*factorized*</u>.

With these preparations we return now to the purification problem as introduced in Section 1.2 . We assume that

(4) the family $(R_i | i\in I)$ is factorized

and that there are given payoff-functions $u_i : \mathsf{X}\, \Omega_j \times \mathsf{X}\, S_j \to \mathbb{R}$ $(i\in I)$ such that

(5) u_i is a sum of tensor products

for every player i. We find therefore μ_i-C-functions $v_i : \Omega_i \times S_i \to \mathbb{R}^n$ for some $n \in \mathbb{N}$ and for every player i with the following properties:

$$R_i(\omega_i, K_1, \ldots, K_N) = R_i^*(\omega_i, \int v_1 \, d\mu_1 \otimes K_1, \ldots, \int v_N \, d\mu_N \otimes K_n)$$

$$(\omega_i \in \Omega_i, \ (K_1, \ldots, K_N) \in \Sigma)$$

for some standard correspondence R_i^* and

$$u_i = \pi_i(v_1, \ldots, v_N)$$

for some componentwise linear function $\pi_i : \mathsf{X}_{i=1}^{N} \mathbb{R}^n \to \mathbb{R}$.

The expected payoff-function $U_i : \mathsf{X}\, \Sigma_j \to \mathbb{R}$ of player i is therefore given by

$$U_i(K_1, \ldots, K_N) = \int u_i \, d(\mu_1 \otimes \ldots \otimes \mu_N) \otimes (K_1 \otimes \ldots \otimes K_N) =$$

$$= \pi_i(\int v_1 \, d\mu_1 \otimes K_1, \ldots, \int v_N \, d\mu_N \otimes K_N) .$$

To simplify notation, for every pair $K, L \in \Sigma$ and $J \subset I$ we define $(K | L_J) := (K_1', \ldots, K_N')$ by $K_i' := K_i$ $(i\in I-J)$ and $K_i' := L_i$ $(i\in J)$. Especially, we have $(K | L_\emptyset) = K$.

The purification problem has now a solution in the following sense:

<u>1.6.3 Lemma:</u> *For given* $K = (K_1,\ldots,K_N) \in \Sigma$ *there is some* $K^* = (\varepsilon_{f_1},\ldots,\varepsilon_{f_N}) \in \Sigma$ *with pure strategies* ε_{f_i} $(i\in I)$ *satisfying*

$$f_i(\omega_i) \in \operatorname{spt} K_i(\omega_i,.) \quad \mu_i\text{-a.e.}\,,$$

$$R_i(.,K) = R_i(.,K^*)$$

and

$$U_i(K|L_J) = U_i(K^*|L_J) \quad (L\in\Sigma,\ J\subset I)$$

for every $i \in I$.

Proof: As an application of Corollary 1.3.4 shows, there are pure strategies ε_{f_i} $(i\in I)$ such that

$$\text{(i)} \qquad \int v_i \, d\mu_i \otimes K_i = \int v_i \, d\mu_i \otimes \varepsilon_{f_i}$$

and

$$\text{(ii)} \qquad f_i(\omega_i) \in \operatorname{spt} K_i(\omega_i,.) \quad \mu_i\text{-a.e.} \quad (i\in I)\,.$$

By (i), $K^* := (\varepsilon_{f_1},\ldots,\varepsilon_{f_N})$ satisfies

$$\text{(iii)} \qquad R_i(.,K) = R_i(.,K^*)$$

for every player i. Moreover, if we set $K_i' := K_i$, $K_i'' = \varepsilon_{f_i}$ $(i\in I-J)$ and $K_i' := K_i'' := L_i$ $(i\in J)$ for given $L \in \Sigma$ and $J \subset I$, we derive from (i) the equation

$$\begin{aligned} U_i(K|L_J) &= \pi_i\Big(\int v_i \, d\mu_1\otimes K_1',\ldots,\int v_N \, d\mu_N\otimes K_N'\Big) = \\ \text{(iv)} \qquad &= \pi_i\Big(\int v_1 \, d\mu_1\otimes K_1'',\ldots,\int v_N \, d\mu_N\otimes K_N''\Big) = \\ &= U_i(K^*|L_J)\,. \end{aligned}$$

Relations (ii)-(iv) result in the statement of the lemma. □

Considering now again the corresponding PRFs $v_i : \Sigma \times \Sigma_i \to \mathbb{R}$ defined by

$$v_i(K,L) := U_i(K|L_i) - U_i(K) \qquad (K \in \Sigma,\ L_i \in \Sigma_i)$$

for player i, we observe that every $K \in \Sigma$ may be replaced by some $K^* \in \Sigma$ consisting of pure strategies and satisfying

$$v_i(K,.) = v_i(K^*,.) .$$

Therefore, K^* is a purification in the sense of 1.2(P). Moreover, the game has an equilibrium in pure strategies.

<u>1.6.4 Theorem:</u> *On condition that assumptions (1)-(5) are satisfied, the game has a pure strategy equilibrium point.*

Proof: By Theorem 1.5.7, the game has an equilibrium point $\overline{K} = (\overline{K}_1,\ldots,\overline{K}_N)$. Lemma 1.6.3 yields a strategy N-tuple $K^* = (\varepsilon_{f_1},\ldots,\varepsilon_{f_N})$ with pure strategies ε_{f_i} such that

(i) $\qquad f_i(\omega_i) \in \operatorname{spt} \overline{K}_i(\omega_i,.) \qquad \mu_i\text{-a.e.}$

(ii) $\qquad R_i(.,\overline{K}) = R_i(.,K^*) \qquad (i \in I)$.

(iii) $\qquad U_i(\overline{K}|L_J) = U_i(K^*|L_J) \qquad (L \in \Sigma, J \subset I)$

Since $\overline{K}$ is an equilibrium point, we have

(iv) $\qquad \operatorname{spt} \overline{K}_i(\omega_i,.) \subset R_i(\omega_i,\overline{K}) \qquad \mu_i\text{-a.e.}$

and

(v) $\qquad U_i(\overline{K}|L_i) \leq U_i(\overline{K}) \qquad$ for every $L_i \in \Sigma_i$ with $\operatorname{spt} L_i \subset R_i(.,\overline{K}) \quad \mu_i\text{-a.e.}$

Relations (i), (ii) and (iv) show that

(vi) $\qquad f_i(\omega_i) \in R_i(\omega_i, K^*) \quad \mu_i\text{-a.e.}$,

(ii), (iii) and (v) yield

(vii) $\qquad U_i(K^*|L_i) = U_i(\bar{K}|L_i) \leq U_i(\bar{K}) = U_i(K^*)$

for every $L_i \in \Sigma_i$ with spt $L_i \subset R_i(.,K^*)$ μ_i-a.e.

On account of (vi) and (vii) K^* is a pure strategy equilibrium point. □

The claims of this section are for instance satisfied in the following situation.

<u>1.6.5 Example:</u> Let the decision spaces S_i $(i\in I)$ be finite; and let the constraints be given by

$$R_i(\omega_i, K) = S_i \qquad (\omega_i \in \Omega_i;\ K \in \Sigma) ;$$

hence, they are inefficient. We set

$$S := \bigtimes_{i\in I} S_i ,$$

and assume that the payoff-function of player i does not depend on the informations of his counter-players. Let therefore be given payoff-functions

$$u_i : \Omega_i \times S \to \mathbb{R} .$$

We claim that all functions $u_i(.,s) : \Omega_i \to \mathbb{R}$ $(s\in S)$ are μ_i-integrable. Define now

$$u_{i,s}(\omega_i) := u_i(\omega_i, s) \qquad (\omega_i \in \Omega_i)$$

$$\delta_{i,s}(t_i) := \begin{cases} 1 & \text{if } s_i = t_i \\ 0 & \text{if } s_i \neq t_i \end{cases} \qquad (t_i \in S_i)$$

for every $s = (s_1 \ldots, s_N) \in S$. Thus, we obtain for every $i \in I$

$$u_i(\omega_i,t) = \sum_{s\in S} u_{i,s}(\omega_i) \prod_{j=1}^{N} \delta_{j,s}(t_j)$$

$$(\omega_i\in\Omega_i,\ t=(t_1,\ldots,t_N)\in S)\ .$$

This formula shows that u_i is a sum of tensor products. Therefore, Lemma 1.6.3 and Theorem 1.6.4 apply to this special case. Games with this type of payoffs have been treated by Radner and Rosenthal (1982).

The next example is not of the type of the previous example. It is related to a military conflict - as many applications of noncooperative game theory.

1.6.6 Example: In a military conflict situation two hostile rocket launching bases are opposed to each other. Every base is equipped with n missiles, each of them bearing g warheads. Every warhead is directed to some rocket of the enemy in such a way that all rockets are threatened by the same number of warheads. The probability of strike is given by a number $\omega \in (0,1)$ for every warhead. Each of both counter-players does not know precisely the probability of strike for the hostile warheads. But, there is some probability distribution μ on $\mathcal{B}((0,1))$ which gives an estimation of the unknown parameter. When both bases simultaneously fire at each other with a certain number s_1 and s_2 of rockets, resp., the number of intact missiles on the first base will be given by

$$(n-s_1)(1-\omega_2)^{gs_2/n} ,$$

provided the probability of strike for the warheads of the second base is $\omega_2 \in (0,1)$. We rather take the interval $[0,n]$ as decision space than $\{0,\ldots,n\}$, because this simplifies calculations. Analogously, the number of intact missiles on the second base amounts to

$$(n-s_2)(1-\omega_1)^{gs_1/n} .$$

The degree of success is therefore given by the payoff-function u_1 with

$$u_1(\omega_1,\omega_2,s_1,s_2) = (n-s_1)(1-\omega_2)^{gs_2/n} - (n-s_2)(1-\omega_1)^{gs_1/n}$$

for the first player. The payoff-function of the counter-player is $u_2 = -u_1$.

Thus, we obtain a typical game with incomplete information - without constraints -, when $((0,1),\mathcal{B}((0,1)),\mu)$ is taken as information space, $([0,n],\mathcal{B}([0,1])$ as decision space of each player and u_1,u_2 as payoff-functions. Obviously, u_1 and u_2 are sums of tensor products. Therefore, we expect the game to have an equilibrium in pure strategies. For reasons of symmetry we suggest that both players will use the same strategy $f:(0,1) \to [0,n]$.

Let

$$p := \int (1-\omega)^{gf(\omega)/n} \, d\mu(\omega)$$

and

$$q := n - \int f \, d\mu .$$

For given probability of strike $\omega \in (0,1)$ decision $s \in [0,1]$ is optimal for player 1, iff s maximizes the expression

$$v(\omega,s') := (n-s')p - q(1-\omega)^{gs'/n}$$

w.r.t. all $s' \in [0,n]$. We obtain

(i) $$\frac{\partial}{\partial s'} v(\omega,s') = - p - q\frac{g}{n}[\ln(1-\omega)](1-\omega)^{gs'/n} .$$

Hence, f must satisfy

$$\frac{\partial}{\partial s'} v(\omega, f(\omega)) = 0 ,$$

or we must have $f(\omega) = 0$ and $f(\omega) = n$, resp. From (i) we derive

$$f(\omega) = \max(0, \min(n, \frac{n}{g \ln(1-\omega)} (c - \ln \ln \frac{1}{1-\omega})))$$

for suitable $c \in \mathbb{R}$. A numerical evaluation of function f for $n = 500$ and $g = 3$ results in the following figure:

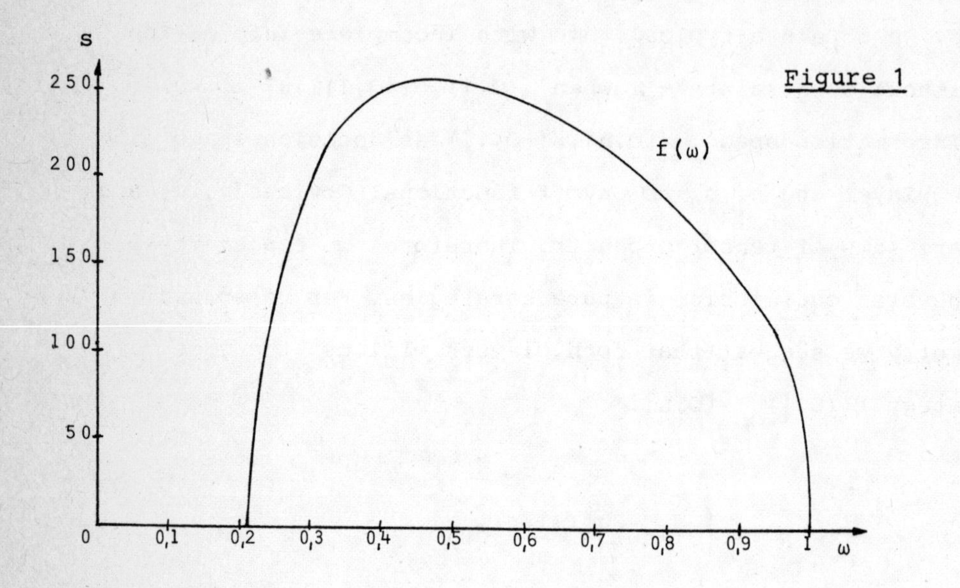

Figure 1

The probability measure μ is concentrated onto a suitable neighbourhood of 0,5 .

Though the model is rather rough, it gives some insight in optimal decision making in military conflict situations of this type. If the strike probability is too low, it is better to do nothing. From a certain limit on, some missiles should be used, but even in case $\omega = 0{,}8$ a rough estimate shows that only 106 missiles remain to the first player while the counter-player has 36 intact missiles.

1.7 Determining approximately payoff-equivalent pure strategies

Most of the results presented until now which are directly related to the purification problem, are essentially based on statement 1.3.4 on existence of payoff-equivalent pure strategies. The proof of this result is not constructive and gives therefore no lead for an explicite determination of such strategies. In the sequel, we will derive a weaker result than 1.3.4, which has the advantage to provide a method to determine approximately payoff-equivalent pure strategies. The idea of proof is due to Aumann et al. (1983). Nevertheless, the result presented in their paper cannot be directly applied to the given situation, since it concerns only finite decision spaces.

1.7.1 Theorem: *Let* $(\Omega, P(\Omega), \mu)$ *be a discrete information space, let* $(S, \mathcal{S})$ *be a decision space, and let* $u : \Omega \times S \to \mathbb{R}^n_+$ *be a measurable function such that a* μ*-integrable function* $g : \Omega \to \mathbb{R}$ *with*

$$\|u(\omega,s)\|^2 \leq g(\omega) \qquad (\omega \in \Omega,\ s \in S)$$

exists. Moreover, let be given a strategy K*, a probability space* $(T, \mathcal{T}, P)$ *and a random variable* $Z : T \times \Omega \to S$ *with the following properties:*

(1) *The family* $(Z(.,\omega) \mid \omega \in \Omega)$ *is independent.*

(2) *For every* $\omega \in \Omega$ *the distribution* $P_{Z(.,\omega)}$ *of* $Z(.,\omega)$ *w.r.t.* P *coincides with* $K(\omega,.)$.

Then, we obtain the estimation

$$E_P\left(\left\|\int u \, d\mu \otimes \varepsilon_{Z(t,.)} - \int u \, d\mu \otimes K\right\|^2\right) \leq \sup_{\omega \in \Omega} \mu(\omega) \int g \, d\mu .$$

Proof: By an application of Fubini's Theorem, we obtain

$$\int\left(\int u \, d\mu\otimes\varepsilon_{Z(t,.)}\right) dP(t) = \int\left(\int u(\omega,Z(t,\omega)) \, d\mu(\omega)\right) dP(t) =$$

$$= \int\left(\int u(\omega,Z(t,\omega)) \, dP(t)\right) d\mu(\omega) =$$

$$\text{(i)} \qquad = \int\left(\int u(\omega,s) \, dP_{Z(.,\omega)}(s)\right) d\mu(\omega) =$$

$$= \int\left(\int u(\omega,s) \, K(\omega,ds)\right) d\mu(\omega) =$$

$$= \int u \, d\mu\otimes K \, .$$

Together with the valuation

$$\sum_{i=1}^{n} V_P(u_i(\omega,Z(.,\omega))) \le \sum_{i=1}^{n} E_P(u_i^2(\omega,Z(.,\omega))) =$$

$$\text{(ii)} \qquad = E_P(\|u(\omega,Z(.,\omega))\|^2) \le g(\omega) \qquad (\omega\in\Omega)$$

we calculate

$$E_P(\|\int u \, d\mu\otimes\varepsilon_{Z(t,.)} - \int u \, d\mu\otimes K\|^2) =$$

$$= \sum_{i=1}^{n} E_P\left(\int u_i \, d\mu\otimes\varepsilon_{Z(t,.)} - \int u_i \, d\mu\otimes K\right)^2 =$$

$$\underset{(i)}{=} \sum_{i=1}^{n} V_P\left(\int u_i \, d\mu\otimes\varepsilon_{Z(t,.)}\right) =$$

$$= \sum_{i=1}^{n} V_P\left(\sum_{\omega\in\Omega} u_i(\omega,Z(t,\omega)) \, \mu(\omega)\right) =$$

$$\underset{(*)}{=} \sum_{i=1}^{n} \left(\sum_{\omega\in\Omega} \mu^2(\omega) \, V_P(u_i(\omega,Z(t,\omega)))\right) \le$$

$$\le (\sup_{\omega\in\Omega} \mu(\omega)) \sum_{\omega\in\Omega} \left[\mu(\omega) \sum_{i=1}^{n} V_P(u_i(\omega,Z(t,\omega)))\right] \le$$

$$\le (\sup_{\omega\in\Omega} \mu(\omega)) \sum \mu(\omega) \, g(\omega) = (\sup_{\omega\in\Omega} \mu(\omega)) \int g \, d\mu \, ,$$

where equation (*) results from the independence of the family $(Z(.,\omega)|\omega\in\Omega)$ and the generalized equation of Bienaimé (A.1.12). □

It deserves to be remarked that a random variable Z with properties (1) and (2) in Theorem 1.7.1 always exists. To this end take

$$(T,\mathcal{T},P) := \bigotimes_{\omega\in\Omega} (S,\mathcal{S},K(\omega,.))$$

and choose for $Z(.,\omega)$ the projection on the coordinate associated with ω. From Theorem 1.7.1 and Markoff's inequality one derives the estimate

$$\begin{aligned} P\{t\in T \,|\, \|\int u \, d\mu\otimes\varepsilon_{Z(t,.)} - \int u \, d\mu\otimes K\|^2 \geq \delta\} &\leq \\ (1) \qquad \leq \frac{1}{\delta} (\sup_{\omega\in\Omega} u(\omega)) \int g \, d\mu \, . \end{aligned}$$

This inequality shows that a pure strategy $\varepsilon_{Z(t,.)}$ with

$$(2) \qquad \|\int u \, d\mu\otimes\varepsilon_{Z(t,.)} - \int u \, d\mu\otimes K\|^2 < \delta$$

exists, whenever the upper bound in (1) becomes smaller than 1. Repeated sampling from distribution P ends therefore up in a pure strategy $\varepsilon_{Z(t,.)}$ satisfying (2) with a probability growing with the sample size. Since $P_{Z(.,\omega)} = K(\omega,.)$ for every $\omega\in\Omega$, an $s = Z(t,\omega) \in S$ is generated by each sample $t \in T$ according to the distribution $K(\omega,.)$. This process is nothing else than playing the mixed strategy K.

For the present, since Theorem 1.7.1 is related to a discrete information space, the nonatomic case is excluded. On the other hand, estimation (1) shows that δ may be chosen very small, if the space $(\Omega, \mathcal{P}(\Omega), \mu)$ approaches to the nonatomic case, i.e., if $\sup_{\omega\in\Omega} \mu(\omega)$ is sufficiently small. In this case, every pure strategy satisfying (2) is almost payoff-equivalent. The following lemma - especially its proof - shows how this approximation can be managed.

<u>1.7.2 Lemma:</u> *Let* $(\Omega, \mathcal{A}, \mu)$ *be a nonatomic, Polish and locally compact information space. Moreover, let* $(S, \mathcal{S})$ *be a compact decision space, and let* $u : \Omega\times S \to \mathbb{R}^n$ *be a* μ*-C-function. Then, for every* $\varepsilon>0$ *there is a finite measurable partition* $(\Omega_i \mid i=0,\dots,r)$ *of* Ω *with the following two properties:*

(1) $0 < \mu(\Omega_i) \leq \varepsilon \quad (i=0,\dots,r)$

(2) *Let the measure* $\bar{\mu}$ *on* $\{0,\dots,r\}$ *be given by* $\bar{\mu}(i) = \mu(\Omega_i)$. *Let* $\bar{u} : \{0,\dots,r\}\times S \to \mathbb{R}^n$ *be defined by*

$$\bar{u}(i,s) := \frac{1}{\mu(\Omega_i)} \int_{\Omega_i} u(\omega,s)\, d\mu \quad (i=0,\dots,r;\ s\in S).$$

Then, for every strategy K *the strategy* $\bar{K}$ *defined by*

$$\bar{K}(i,A) := \frac{1}{\mu(\Omega_i)} \int_{\Omega_i} K(\omega,A)\, d\mu \quad (i=0,\dots,r;\ A\in\mathcal{S})$$

satisfies

$$\left\| \int \bar{u}\, d\bar{\mu}\otimes\bar{K} - \int u\, d\mu\otimes K \right\| \leq \varepsilon .$$

Proof: Since u is a μ-C-function, there is a μ-integrable function $g : \Omega \to \mathbb{R}$ which dominates $\|u\|$. As a consequence of the regularity of μ and A.1.9, there exists a compact set $C \subset \Omega$ such that $\mu(\Omega-C) \leq \varepsilon$,

(i) $$\int_{\Omega-C} g\, d\mu \leq \frac{\varepsilon}{4},$$

and such that u is uniformly continuous on $C\times S$. Therefore, one can find a partition of C into measurable sets $\Omega_1,\dots,\Omega_r$ with the property

$$\text{(ii)} \qquad \|u(\omega,s) - u(\omega',s)\| \leq \frac{\varepsilon}{2} \qquad (\omega,\omega'\in\Omega_i,\ s\in S)$$

for every $i = 1,\dots,r$. The sets $\Omega_1,\dots,\Omega_r$ can be chosen so as to satisfy

$$\mu(\Omega_i) \leq \varepsilon \qquad (i=1,\dots,r) .$$

Define $\Omega_o := \Omega - C$. W.l.o.g. we may assume $\mu(\Omega_i) > 0$ $(i=0,\dots,r)$. We define now the measure $\overline{\mu}$, the function $\overline{u}$ and the strategy $\overline{K}$ - for a given strategy K - according to (2); and we introduce the function $\tilde{u}$ and the strategy $\tilde{K}$ by

$$\tilde{u}(\omega,.) = \overline{u}(i,.); \quad \tilde{K}(\omega,.) := \overline{K}(i,.) \qquad (\omega\in\Omega_i;\ i=0,\dots,r) .$$

Then, we obtain

$$\text{(iii)} \qquad \int \overline{u}\, d\overline{\mu}\otimes\overline{K} = \int \tilde{u}\, d\mu\otimes\tilde{K} = \int \tilde{u}\, d\mu\otimes K .$$

From (ii) we derive the estimate

$$\text{(iv)} \qquad \|\tilde{u}(\omega,s) - u(\omega,s)\| \leq \frac{\varepsilon}{2}$$

for every $\omega \in \Omega_i$ for some $i \in \{1,\dots,r\}$; and from (i) we derive for every $\omega \in \Omega_o$

$$\text{(v)} \qquad \|\tilde{u}(\omega,s)\| \leq \frac{1}{\mu(\Omega_o)} \int_{\Omega_o} \|u(\omega,s)\|\, d\mu \leq \frac{\varepsilon}{4\mu(\Omega_o)} .$$

Relations (i), (iii), (iv) and (v) result in the valuation

$$\left\|\int \overline{u}\, d\overline{\mu}\otimes\overline{K} - \int u\, d\mu\otimes K\right\| \leq$$

$$\leq \int_{\Omega_o\times S} \|\tilde{u} - u\|\, d\mu\otimes K + \sum_{i=1}^{r} \int_{\Omega_i\times S} \|\tilde{u} - u\|\, d\mu\otimes K \leq$$

$$\leq \frac{\varepsilon}{4} + \int_{\Omega_o} g \, d\mu + \frac{\varepsilon}{2} \sum_{i=1}^{r} \mu(\Omega_i) = \varepsilon . \qquad \square$$

§ 2 A MARKET GAME AS A GAME WITH INCOMPLETE INFORMATION

2.1 A model of a market game with a continuum of traders

In the sequel, we investigate a market game. Unless the model developed by Hildenbrand (1974) we assume that preferences of traders are given by payoff-functions. Nevertheless, we allow the payoffs to be price-dependent. An extended version of such a market game including production has been introduced by Greenberg et al. (1977). With the model to be introduced in the sequel we do not set great store by most generality, we are rather interested in the connection to the theory of games with incomplete information.

The model is related to the situation of a comprehensive group of agents trading in commodities at a market. The group of traders is represented by a probability space (Ω, A, μ). The σ-field A has the meaning of the set of possible coalitions, whereas the measure μ gives the fraction of each coalition w.r.t. the whole group. The number of types of commodities offered at the market is assumed to be finite. The types of commodities are enumerated by $1, \ldots, n$. Every vector $x \in \mathbb{R}^n_+$ represents a bundle of commodities; the k-th component x_k gives the quantity of commodity k. The set of all possible bundles is assumed to be given by a compact set $S \subset \mathbb{R}^n_+$. Let the elements of S be column vectors - for technical reasons. Every player has some initial endowment of commodities, which are given by a measurable function

$$a : \Omega \to S .$$

The set of price vectors for the different types of commodities is represented by the unit simplex Δ. The k-th component of $p \in \Delta$ means the price of one unit of commodity k. Since we are only interested in relations of prices of different commodities, it suffices to consider price vectors from the unit simplex. Preferences of traders are given by a payoff-function

$$u : \Omega \times S \times \Delta \to \mathbb{R} .$$

We will require that u is a μ-C-function, i.e.

$$u(.,x,p) \text{ is } A\text{-measurable} \quad (x \in S, p \in \Delta) ,$$

$$u(\omega,.,.) \text{ is continuous on } S \times \Delta \quad (\omega \in \Omega) ,$$

and a μ-integrable function $g : \Omega \to \mathbb{R}$ exists such that

$$|u(\omega,x,p)| \leq g(\omega) \quad (\omega \in \Omega,\ x \in S,\ p \in \Delta) .$$

Since the initial endowment $a(\omega)$ of trader $\omega \in \Omega$ is usually not optimal within the set S w.r.t.the payoff-function $u(\omega,.,p)$ for given prices p, there is some incentive for traders to exchange their commodities for those of other traders. Demand and supply will influence prices of commodities at the market. Prices will again determine the set of bundles which traders may buy. The set of all those bundles for trader $\omega \in \Omega$ is given by

$$R(\omega,p) := \{x \in S \mid px \leqq pa(\omega)\} \quad (p \in \Delta) .$$

We assume for technical reasons that Δ consists of line vectors. The set $R(\omega,p)$ is called the <u>budget set</u> of trader ω w.r.t. price vector p. The following solution concept should be regarded as a disposition how a fair distribution of commodities and corresponding prices might be adjusted at the market.

<u>2.1.1 Definition:</u> *Every measurable function* $X : \Omega \to S$ *is called an* <u>*allocation*</u>. *Every allocation* X, *for which a price vector* $p \in \Delta$ *exists such that*

(1) $X(\omega) \in R(\omega,p)$ μ-a.e. ,

(2) $u(\omega,X(\omega),p) = \max_{x \in R(\omega,p)} u(\omega,x,p)$ μ-a.e. ,

(3) $\int X \, d\mu \leq \int a \, d\mu$,

holds, is called a <u>*Walras-allocation*</u>. *The pair* (X,p) *is called a* <u>*Walras-equilibrium*</u> .

Condition (1) of Definition 2.1.1 means that almost every player can only command such bundles which he can finance, (2) means, that trader ω prefers no other bundle within his budget set, and (3) means that allocation X can be realized at the market.

This type of market game can now be brought into the form of a game with incomplete information. The probabitity space (Ω,A,μ) becomes the information space of the group of traders Ω, S becomes the decision space and R becomes the constraint correspondence. As an artificial counter player we introduce a price-player (without information space). His decision space is given by $\{1,\ldots,n\}$. Since every probability distribution on $\{1,\ldots,n\}$ is in a one-to-one correspondence with a vector from the unit simplex Δ, the mixed strategies of the price-player are given by Δ. The strategy set of the trader group is the set

$$\Sigma := \{K | K : \Omega \times \mathcal{B}(S) \to [0,1] \text{ is a stoch. kernel}\}$$

endowed with the corresponding weak topology. Every measurable function $X : \Omega \to S$ - i.e. every allocation - may therefore be regarded as a pure strategy.

We assume that (Ω, A, μ) is a locally compact and Polish space. Moreover, we introduce the preference representation functions $v_1 : (\Sigma \times \Delta) \times \Sigma \to \mathbb{R}$ and $v_2 : (\Sigma \times \Delta) \times \Delta \to \mathbb{R}$ by

$$v_1(K,p;L) := \int u(.,p) \, d\mu \otimes L - \int u(.,p) \, d\mu \otimes K$$

$$v_2(K,p;q) := (p-q) \left(\int a \, d\mu - \int id_S \, d\mu \otimes K\right)$$

$$(K,L \in \Sigma;\ p,q \in \Delta) .$$

Thus, we obtain a game with incomplete information. Continuity of v_1 and v_2 is a consequence of 1.4.5 .

These considerations suggest the following extension of Definition 2.1.1 .

<u>2.1.2 Definition:</u> *Every strategy* $K \in \Sigma$ *is called a* <u>*randomized allocation*</u>. *Every* $K \in \Sigma$, *for which a* $p \in \Delta$ *exists such that*

(1) $K(\omega, R(\omega,p)) = 1 \quad \mu\text{-a.e.}$,

(2) $u(\omega,.,p) = \max_{x \in R(\omega,p)} u(\omega,x,p) \quad K(\omega,.)\text{-a.e. for } \mu\text{-a.e. } \omega \in \Omega$,

(3) $\int \left(\int x \, K(\omega,dx) \right) d\mu(\omega) \leq \int a \, d\mu$

holds, is called a <u>*randomized Walras-allocation*</u>.

The pair (K,p) *is called a* <u>*randomized Walras-equilibrium*</u> *(abbreviated:* <u>*r-allocation, r-Walras-allocation, r-Walras-equilibrium*</u>*)* .

Especially, every r-Walras-equilibrium (K,p) with a pure strategy K is a Walras-equilibrium in the sense of Definition 2.1.1 . By means of the purification theory as developed in Section 1.3 we obtain the result that every r-Walras-equilibrium can be purified. But first, we will investigate the budget sets more precisely.

<u>2.1.3 Lemma:</u> *For every* $p \in \Delta$ *the correspondence* $R(.,p)$ *is* $\mathcal{A}$-$\mathcal{B}(C(S))$*-measurable.*

Proof: For given $p \in \Delta$ we have

$$\{\omega\in\Omega \mid R(\omega,p) \cap F \neq \emptyset\} = \{\omega\in\Omega \mid \min_{x\in F} px \leq pa(\omega)\} \in \mathcal{A}$$

for every compact set $F \subset S$. Since S is compact, we obtain therefore

$$\{\omega\in\Omega \mid R(\omega,p) \subset U\} \in \mathcal{A}$$

for every open set $U \subset S$. From this and A.2.6 we conclude that $R(.,p)$ is $\mathcal{A}$-$\mathcal{B}(C(S))$-measurable. □

The following lemma will be used to show existence of a Walras-equilibrium.

<u>2.1.4 Lemma:</u> *If an r-Walras-equilibrium* (K,p) *exists, then there is an allocation* X *for which* (X,p) *is a Walras-equilibrium.*

Proof: From 1.3.4 we derive that for every r-Walras-equilibrium (K,p) there exists a pure strategy ε_X such that

(i) $$\int u(.,p)\; d\mu\otimes\varepsilon_X = \int u(.,p)\; d\mu\otimes K\,,$$

(ii) $$\int id_S\; d\mu\otimes\varepsilon_X = \int id_S\; d\mu\otimes K$$

and

(iii) $$X(\omega) \in R(\omega,p) \quad \mu\text{-a.e.}$$

These three properties of the allocation X and 1.4.10 show that (X,p) is a Walras-equilibrium. □

A connection between equilibria of the described game with incomplete information and r-Walras-equilibria is subject of the following theorem.

<u>2.1.5 Theorem:</u> *Every equilibrium of the game with incomplete information is an r-Walras-equilibrium of the market game.*

Proof: Let (K,p) be an equilibrium point. Then, the following relations are satisfied:

(i) $$K(\omega,R(\omega,p)) = 1 \quad \mu\text{-a.e.}$$

and

(ii) $$\begin{aligned} &v_1(K,p;L) \leq 0 \quad \text{for every } L \in \Sigma \text{ with } L(\omega,R(\omega,p)) = 1 \quad \mu\text{-a.e.}\,,\\ &v_2(K,p;q) \leq 0 \quad \text{for every } q \in \Delta\,. \end{aligned}$$

From (ii) we derive

(iii) $$\int u(.,p)\; d\mu\otimes L - \int u(.,p)\; d\mu\otimes K = v_1(K,p;L) \leq 0$$

for every $L \in \Sigma$ with $L(\omega,R(\omega,p)) = 1 \quad \mu$-a.e.

and

$$\text{(iv)} \qquad (p-q)\left(\int a\, d\mu - \int\left(\int x\, K(\omega,dx)\right) d\mu\right) =$$

$$= v_2(K,p;q) \leq 0 \qquad (q \in \Delta) .$$

From (i), (iii) and 1.4.10, we get the relation

$$\text{(v)} \qquad u(\omega,.,p) = \max_{x \in R(\omega,p)} u(\omega,x,p) \quad K(\omega,.)\text{-a.e.}$$
$$\text{for } \mu\text{-a.e. } \omega \in \Omega .$$

Moreover, (i) yields

$$p \int x\, K(\omega,dx) = p \int_{R(\omega,p)} x\, K(\omega,dx) \leq pa(\omega)$$
$$\mu\text{-a.e.}$$

Therefore, the estimate

$$p\left(\int a\, d\mu - \int\left(\int x\, K(\omega,dx)\right) d\mu\right) \geq 0$$

holds. Choosing for q all unit vectors in Δ, we conclude from this estimate together with (iv) that

$$\text{(vi)} \qquad \int a\, d\mu \geq \int\left(\int x\, K(\omega,dx)\right) d\mu$$

is satisfied. Relations (i), (v) and (vi) show that (K,p) is an r-Walras-equilibrium. □

By this theorem existence of an r-Walras-equilibrium of the market game is proved, provided the game with incomplete information has an equilibrium. Before such an existence theorem will be established, we first make some preparations.

The set S is called <u>contractible</u>, iff

$$\alpha \in [0,1],\ x \in S \implies \alpha x \in S$$

holds. This property of S ensures a continuity property of the correspondence R.

<u>2.1.6 Lemma:</u> *Let S be contractible; let* $\omega \in \Omega$ *and* $p_o \in \Delta$ *with* $p_o a(\omega) > 0$ *be given. Then, the correspondence* $R(\omega,.)$ *is continuous at* p_o .

Proof: Let (p_n) be a sequence from Δ such that $\lim_{n\to\infty} p_n = p_o$. For every sequence (x_n) with

$$x_n \in R(\omega,p_n)$$

and

$$\lim_{n\to\infty} x_n = x \in S ,$$

we have

$$p_o x = \lim_{n\to\infty} p_n x_n \leq \lim_{n\to\infty} p_n a(\omega) = p_o a(\omega) ;$$

and therefore

$$x \in R(x,p_o) .$$

By A.2.3 this shows upper semicontinuity of $R(\omega,.)$ at p_o .

Let now $x \in R(\omega,p_o)$ be given. Further let $0 \leq \alpha < 1$. Since S is contractible, αx is an element of S ; and the inequalities

$$p_o(\alpha x) \leq p_o x \leq p_o a(\omega)$$

hold, where at least one of them is a strict inequality, because $p_o a(\omega) > 0$. This shows

$$p_n(\alpha x) \leq p_n a(\omega) \quad \text{for almost every } n \in \mathbb{N} ,$$

i.e.

$$\alpha x \in R(\omega,p_n) \quad \text{for almost every } n \in \mathbb{N} .$$

By A.2.3 this proves lower semicontinuity of $R(\omega,.)$ at p_o. □

The following existence theorem bases on these preliminaries.

2.1.7 Theorem: *Let S be contractible, and let a be strictly positive. Then the game with incomplete information has an equilibrium point.*

Proof: From 2.1.3 and 2.1.6 we derive that $R : \Omega\times\Delta \to C(S)$ is a C-constraint correspondence - 2.1.6 applies because a is strictly positive. Therefore, by the existence theorem 1.5.1 an equilibrium point exists. □

Together with 2.1.5 and 2.1.4, this theorem shows that a Walras-equilibrium of the market game exists in case of contractibility of S and strict positivity of a. Insofar 2.1.7 is an existence theorem for a Walras-equilibrium of the market game. The claim of strict positivity of a is surely in general not very realistic, since usually one cannot expect that every player has some quantity of all commodities. In the sequel this claim will be weakened in some sense. The following example shows that the conditions on u have to be strengthened, when a is not strictly positive.

2.1.8 Example: We start with $(\Omega,A,\mu) := ([0,3],\mathcal{B}([0,3]),\frac{1}{3}\lambda)$, $S := [0,3]^2$,

$$a(\omega) := \begin{cases} (1,0)^t & \text{for } \omega \in [0,1) \\ (1,1)^t & \text{for } \omega \in [1,2) \\ (0,1)^t & \text{for } \omega \in [2,3] \end{cases}$$

and

$$u(\omega,x,p) := \begin{cases} x_1 & \text{for } \omega \in [0,2) \\ x_2 & \text{for } \omega \in [2,3] \end{cases} \qquad (x \in S,\ p \in \Delta)\ .$$

Let now X be an allocation, and let $p \in \Delta$ such that

$$X(\omega) \in R(\omega,p) \quad \mu\text{-a.e.}$$

and

$$u(\omega,X(\omega),p) = \max_{x \in R(\omega,p)} u(\omega,x,p) \quad \mu\text{-a.e.}$$

Because of

$$R(\omega,p) = \begin{cases} \{x \in S \mid px \leq p_1\} & \text{for } \omega \in [0,1) \\ \{x \in S \mid px \leq 1\} & \text{for } \omega \in [1,2) \\ \{x \in S \mid px \leq p_2\} & \text{for } \omega \in [2,3] \end{cases} \qquad (p \in \Delta)\ ,$$

X satisfies for every $p \in \Delta$ with $0 < p_1 < 1$

$$X_1(\omega) = \begin{cases} 1 & \text{for } \omega \in [0,1) \\ \min\{3, \frac{1}{p_1}\} & \text{for } \omega \in [1,2) \end{cases} \qquad \mu\text{-a.e.}\ ,$$

in case $p_1 = 0$, i.e. $p = (0,1)$, we have

$$X_1(\omega) = 3 \quad \mu\text{-a.e. within } [0,2)\ .$$

Therefore, in each case

$$\int X_1\, d\mu > \frac{2}{3} = \int a_1\, d\mu\ .$$

Hence, only for $p = (1,0)$ a Walras-equilibrium might exist. But in this case we have

$$X_2(\omega) = 3$$

μ-a.e. within $[2,3]$, and therefore

$$\int X_2 \, d\mu \geqq 1 > \frac{2}{3} = \int a_2 \, d\mu .$$

This shows that the market game cannot have a Walras-equilibrium.

From now on, we assume S to be a rectangle in $\mathbb{R}^n_+$, i.e. $S = [0,c]$ for some $c \in \mathbb{R}^n_+$. In particular, the so given S is contractible. The payoff-function u is called <u>monotonic</u> (w.r.t. the semiorder of $\mathbb{R}^n$), iff

$$x,y \in S, \; x \lneqq y \implies u(\omega,x,p) < u(\omega,y,p)$$

holds for every $\omega \in \Omega$ and $p \in \Delta$. The initial allocation a is called <u>balanced</u>, if

$$a(\omega) \neq 0 \quad (\omega \in \Omega), \quad \int a \, d\mu > 0$$

holds, and if the coalitions

$$\Omega_j = \{\omega \in \Omega \mid a_j(\omega) > 0\} \qquad (j=1,\ldots,n)$$

satisfy the relations

$$c_i \mu(\Omega_j) > \int a_i \, d\mu \qquad (i,j = 1,\ldots,n) .$$

Balancedness of a means therefore that the maximal demand c_i of commodity i cannot be met for any of the coalitions Ω_j.

By an n-1-dimensional simplex in $\mathbb{R}^n$ we understand the convex hull of n linear independent vectors from $\mathbb{R}^n$. We first obtain the following result.

<u>2.1.9 Lemma:</u> *Suppose that the strategies of the price-player are confined to some n-1-dimensional simplex* $\Delta^* \subset \Delta$. *Then, every equilibrium point* (K,p) *of the confined game with*

$p \in \operatorname{int} \Delta^*$ *is an equilibrium point of the original game.*

The proof becomes obvious by an inspection of the equilibrium conditions. The second condition

$$v_2(K,p;q) \leq 0 \quad (q \in \Delta)$$

is satisfied, because it depends only on the differences $p-q$ $(q \in \Delta)$, which are multiples of the differences $p-q$ $(q \in \Delta^*)$.

The next lemma shows that in case of monotonicity of u and balancedness of a the demand for commodities having price zero is not met by the supply.

<u>2.1.10 Lemma:</u> *Let* $((K^{(k)},p^{(k)}) \mid k \in \mathbb{N})$ *be a sequence from* $\Sigma \times \Delta$, *and let* $(K,p) = \lim_{k\to\infty} (K^{(k)},p^{(k)})$. *For some* $i \in \{1,\ldots,n\}$, *we assume that* $p_i = 0$. *Further, let* $K^{(k)}(\omega,R(\omega,p^{(k)})) = 1$ μ-a.e., *and let*

$$u(\omega,.,p^{(k)}) = \max_{x \in R(\omega,p^{(k)})} u(\omega,x,p^{(k)}) \quad K^{(k)}(\omega,.)\text{-a.e.}$$

$$\text{for } \mu\text{-a.e. } \omega \in \Omega .$$

Then, this implies in case of monotonicity of u *and balancedness of* a *that*

$$\int \left(\int x_i \, K(\omega,dx) \right) d\mu > \int a_i \, d\mu$$

holds.

Proof: Since $p \in \Delta$, there is some $j \in \{1,\ldots,n\}$ with $p_j > 0$. Define now

$$U(L,q) := \int_{\Omega_j \times S} u(.,q) \, d\mu \otimes L \quad (L \in \Sigma, q \in \Delta) .$$

Then, as a consequence of 1.4.5, U is continuous on $\Sigma \times \Delta$. Since $a_j(\omega) > 0$ $(\omega \in \Omega_j)$, we have $pa(\omega) > 0$ $(\omega \in \Omega_j)$. Therefore, the correspondence $R(\omega,.)$ is continuous at p for every $\omega \in \Omega_j$, as 2.1.6 shows. Define now R^* by

$$R^*(\omega,.) := \begin{cases} R(\omega,.) & \text{for } \omega \in \Omega_j \\ S & \text{for } \omega \in \Omega - \Omega_j . \end{cases}$$

Thus, $R^*(\omega,.)$ is continuous at p for every $\omega \in \Omega$. Setting

$$\Sigma_q := \{L \in \Sigma \mid L(\omega, R^*(\omega,q)) = 1 \quad \mu\text{-a.e.}\} \qquad (q \in \Delta)$$

we conclude from $K^{(k)} \in \Sigma_{p^{(k)}}$ $(k \in \mathbb{N})$, from

$$U(K^{(k)}, p^{(k)}) = \max_{L \in \Sigma_{p^{(k)}}} U(L, p^{(k)}) \qquad (k \in \mathbb{N})$$

and 1.4.9 together with A.2.4 that $K \in \Sigma_p$

and

$$U(K,p) = \max_{L \in \Sigma_p} U(L,p)$$

holds. Hence, from 1.4.10, we get

$$\text{(i)} \qquad u(\omega,.,p) = \max_{x \in R(\omega,p)} u(\omega,x,p) \quad K(\omega,.)\text{-a.e.}$$

for μ-a.e. $\omega \in \Omega_j$. Since u is monotonic and $p_i = 0$, every $\omega \in \Omega$ and $y \in R(\omega,p) = \{x \in [0,c] \mid px \le pa(\omega)\}$ satisfy the implication

$$u(\omega,y,p) = \max_{x \in R(\omega,p)} u(\omega,x,p) \implies y_i = c_i .$$

Together with (i), this shows

$$K(\omega, \{x \in [0,c] \mid x_i = c_i\}) = 1 \quad \text{for } \mu\text{-a.e. } \omega \in \Omega_j .$$

Therefore, the estimation

$$\int \left(\int x_i \; K(\omega, dx) \right) d\mu \geq \int_{\Omega_j} \left(\int x_i \; K(\omega, dx) \right) d\mu =$$

$$= \int_{\Omega_j} c_i \; d\mu = c_i \mu(\Omega_j)$$

holds, which proves the lemma, since a is balanced. □

By means of 2.1.9 and 2.1.10, a second result on existence of equilibrium may be proved.

<u>2.1.11 Theorem:</u> *If* u *is monotonic, and if* a *is balanced, then the game has an equilibrium point.*

Proof: Let u be monotonic, and let a be balanced. Let $(\Delta^{(k)})$ be an increasing sequence of n-1-dimensional simplices $\Delta^{(k)} \subset \text{int}\,\Delta$ such that

$$\Delta^{(k)} \uparrow \text{int}\,\Delta\,.$$

Further, let $k \in \mathbb{N}$. Since $\Delta^{(k)} \subset \text{int}\,\Delta$, we have $p > 0$ for $p \in \Delta^{(k)}$ and $\omega \in \Omega$. From 2.1.3 and 2.1.6 we conclude, that $R : \Omega \times \Delta^{(k)} \to C(S)$ is a C-constraint correspondence. If the price-player confines to strategies from $\Delta^{(k)}$, the confined games satisfies all assumptions of Theorem 1.5.1, and has therefore an equilibrium point $(K^{(k)}, p^{(k)})$. If a $k_o \in \mathbb{N}$ with $p^{(k_o)} \in \text{int}\,\Delta^{(k_o)}$ exists, the theorem is proved by 2.1.9. Otherwise $p^{(k)}$ is a boundary point of $\Delta^{(k)}$ for every $k \in \mathbb{N}$. Since Σ and Δ are compact sets, we may assume that $(K,p) \in \Sigma \times \Delta$ exists such that

$$\lim_{k\to\infty} K^{(k)} = K, \quad \lim_{k\to\infty} p^{(k)} = p\,.$$

Hence, p is a boundary point of Δ and has therefore a component $p_i = 0$. Since $(K^{(k)}, p^{(k)})$ is an equilibrium point of the confined game, the sequence $((K^{(k)}, p^{(k)}) \mid k \in \mathbb{N})$ satisfies the assumptions of 2.1.10. Consequently,

$$\text{(i)} \qquad \int \left(\int x_i \, K(\omega, dx) \right) d\mu > \int a_i \, d\mu$$

holds true. Since $(K^{(k)}, p^{(k)})$ is an equilibrium point of the corresponding game, $p^{(k)}$ satisfies

$$(p^{(k)} - q) \left(\int a \, d\mu - \int \left(\int x \, K^{(k)}(\omega, dx) \right) d\mu \right) \leq 0 \qquad (q \in \Delta^{(k)})$$

and $K^{(k)}$ satisfies $K^{(k)}(\omega, R(\omega, p^{(k)})) = 1$ μ-a.e.

The latter implies

$$p^{(k)} \left(\int a \, d\mu - \int \left(\int x \, K^{(k)}(\omega, dx) \right) d\mu \right) \geq 0 .$$

For continuity reasons we have therefore

$$(p - q) \left(\int a \, d\mu - \int \left(\int x \, K(\omega, dx) \right) d\mu \right) \leq 0 \qquad (q \in \Delta)$$

and

$$p \left(\int a \, d\mu - \int \left(\int x \, K(\omega, dx) \right) d\mu \right) \geq 0 .$$

From this, we conclude that

$$\int a \, d\mu - \int \left(\int x \, K(\omega, dx) \right) d\mu \geq 0$$

holds, which is a contradiction to (i). This proves the theorem. □

Moreover, 2.1.10 shows that the price vector p of an r-Walras-equilibrium (K,p) is strictly positive in the situation discussed here - to this end one has only to consider the sequence $(K^{(k)}, p^{(k)}) := (K,p)$ $(k \in \mathbb{N})$ in 2.1.10. No commodity is therefore offered free of charge at the market.

Theorem 2.1.11 is similar to Theorem 3.1 , Kap. III, in Rosenmüller (1971). The main difference is that preferences of players are allowed to be price-dependent in 2.1.11 . The result differs from that of Greenberg et al. (1977) in so far as assumption (A.5) on the initial endowment a made there is stronger than the assumption of balancedness for a made in this context now.

2.2 <u>On the connection between core, r-core and the set of r-Walras allocations</u>

As in the last part of the previous section we assume that S is a rectangular subset of $\mathbb{R}^n_+$, i.e. $S = [0,c]$ for some $c \in \mathbb{R}^n_+$. We adopt the model developed in 2.1 with the only difference that the payoff-function u does not depend on prices now; i.e. u is defined on $\Omega \times S$. We first introduce a concept of domination on the set Σ of all strategies which allows a completely different solution concept for the market game compared to the concept of an r-Walras allocation.

2.2.1 Definition: *A strategy* $L \in \Sigma$ *is said to* dominate *the strategy* $K \in \Sigma$ stochastically *w.r.t. a coalition* $A \in \mathcal{A}$ *(abbreviated:* $L \succsim_A K$*), iff*

$$L(\omega,\{u(\omega,.) \geqq \alpha\}) \geqq K(\omega,\{u(\omega,.) \geqq \alpha\}) \qquad (\alpha \in \mathbb{R})$$

holds for μ*-almost every* $\omega \in A$.

The strategy $L \in \Sigma$ *is said to* dominate $K \in \Sigma$ strongly stochastically *w.r.t.* $A \in \mathcal{A}$ *(abbreviated:* $L \succ_A K$*), iff* $L \succsim_A K$ *holds but* $K \succsim_A L$ *is not satisfied.*

The set of strategies K such that

$$\int \left(\int x \, K(\omega,dx) \right) d\mu \leqq \int a \, d\mu$$

holds, and such that there exist no strategy L and no set $A \in \mathcal{A}$ *with* $L \succ_A K$ *and*

$$\int_A \left(\int x \, L(\omega,dx) \right) d\mu \leqq \int_A a \, d\mu ,$$

is called the randomized core *of the market game (shortly:* r-core*).*

The set of all allocations X *such that*

$$\int X \, d\mu \leq \int a \, d\mu \ ,$$

for which no allocation Y *and no set* $A \in \mathcal{A}$ *with* $\varepsilon_Y \succ_A \varepsilon_X$ *and*

$$\int_A Y \, d\mu \leq \int_A a \, d\mu$$

exist, is called the <u>*core*</u> *of the market game.*

For two strategies $K, L \in \Sigma$ the property $L \succsim_A K$ means that every player in coalition A achieves every payoff level $\alpha \in \mathbb{R}$ by at least the same probability, when L is used instead of K. If $L \succ_A K$ holds, then coalition A will prefer to use L instead of K. In case K is an element of the r-core, there is no coalition A which can divide its bundle of commodities according to a strategy L in such a way that $L \succ_A K$ holds. Some properties of the stochastic dominance relation as introduced in 2.2.1 are captured in the subsequent remarks.

<u>2.2.2 Remarks:</u> (1) For two allocations X and Y the relation $\varepsilon_Y \succsim_A \varepsilon_X$ is equivalent to the existence of the inequality

$$u(\omega, Y(\omega)) \geq u(\omega, X(\omega))$$

for μ-almost every $\omega \in A$.

(2) For an allocation X and a strategy L the relation $L \succsim_A \varepsilon_X$ holds, iff

$$L(\omega, \{u(\omega, .) \geq u(\omega, X(\omega))\}) = 1$$

is satisfied for μ-almost every $\omega \in A$.

(3) For a strategy K, and an allocation Y the relation $\varepsilon_Y \succsim_A K$ is synonymous with

$$K(\omega,\{u(\omega,.) \leqq u(\omega,Y(\omega))\}) = 1$$

for μ-almost every $\omega \in A$.

The aim of the further investigations is to show the connections between the different solution concepts. First, we establish that the core consists of all pure strategies from the r-core, i.e. all allocations from the r-core.

2.2.3 Lemma: *Let X be an allocation. Then X is an element of the core if and only if* ε_X *belongs to the r-core.*

Proof: Let X be an element of the core. Suppose, there exist a strategy L and a set $A \in \mathcal{A}$ with $L \succ_A \varepsilon_X$ and

$$\text{(i)} \qquad \int_A \Big(\int x \, L(\omega,dx) \Big) \, d\mu \leqq \int_A a \, d\mu \, .$$

Then, we have $L \succsim_A \varepsilon_X$ and therefore by 2.2.2(2)

$$\text{(ii)} \qquad L(\omega,\{u(\omega,.) \geqq u(\omega,X(\omega))\}) = 1 \quad \mu\text{-a.e. within } A \, .$$

On the other hand, $\varepsilon_X \succsim_A L$ cannot be satisfied. Therefore, by 2.2.2(3), a set $B \in \mathcal{A}$ exists with $B \subset A$, $\mu(B)>0$ and

$$L(\omega,\{u(\omega,.) \leqq u(\omega,X(\omega))\}) < 1 \quad (\omega \in B) \, ,$$

the latter being equivalent to

$$\text{(iii)} \qquad L(\omega,\{u(\omega,.) > u(\omega,X(\omega))\}) > 0 \quad (\omega \in B) \, .$$

An application of 1.3.4 to the information space $(A, A \cap \mathcal{A}, \mu|_A)$, to the function $(id_S,u) : A \times S \to \mathbb{R}^{n+1}$, to the correspondence $F : A \to S$ defined by

$$F(\omega) := \{x \in S \mid u(\omega,x) \geq u(\omega,X(\omega))\} \qquad (\omega \in A)$$

and to the strategy L gives a measurable mapping $Y' : A \to S$ with properties

(iv) $$\int_A Y' \, d\mu = \int_A \left(\int x \, L(\omega,dx) \right) d\mu$$

(v) $$\int_A u(\omega,Y'(\omega)) \, d\mu = \int_A \left(\int u(\omega,x) \, L(\omega,dx) \right) d\mu$$

(vi) $Y'(\omega) \in F(\omega)$ μ-a.e. within A.

From (i) and (iv), we derive

(vii) $$\int_A Y' \, d\mu \leq \int_A a \, d\mu .$$

Relations (v), (ii) and (iii) imply

(viii) $$\int_A u(\omega,Y'(\omega)) \, d\mu > \int_A u(\omega,X(\omega)) \, d\mu .$$

Define now the allocation Y by

$$Y(\omega) = \begin{cases} Y'(\omega) & (\omega \in A) \\ X(\omega) & (\omega \in \Omega - A) . \end{cases}$$

Thus, because of (vi), we get

$$u(\omega,Y(\omega)) \geq u(\omega,X(\omega)) \qquad \mu\text{-a.e. within } A,$$

i.e. $\varepsilon_Y \succsim_A \varepsilon_X$, by 2.2.2(1). On the other hand, (viii) shows that we cannot have

$$u(\omega,Y(\omega)) \leq u(\omega,X(\omega)) \qquad \mu\text{-a.e. within } A.$$

Therefore, by 2.2.2(1), $\varepsilon_X \succsim_A \varepsilon_Y$ is not valid. Consequently, $\varepsilon_Y \succ_A \varepsilon_X$ holds, which contradicts the fact, that X was an element of the core. The converse direction of the proof is an immediate consequence of the definitions of the core and r-core. □

Before treating a connection between the set of r-Walras allocations and the r-core, we prove a preparing lemma.

2.2.4 Lemma: *Let* u *be monotonic, let* $p \in \Delta$ *and let* $\omega \in \Omega$. *Then,*

$$\{z \in R(\omega,p) \mid u(\omega,z) < \max_{x \in R(\omega,)} u(\omega,x)\} \supset \{z \in S \mid pz < pa(\omega)\}$$

holds.

Proof: Let be given $z \in S = [0,c]$ such that $pz < pa(\omega)$. Then, we have $v := c - z \underset{\neq}{\geq} 0$; - otherwise we would have $z = c \geqq a(\omega)$, hence $pz \geqq pa(\omega)$. Therefore, we can find some $\alpha \in \mathbb{R}$ with $0 < \alpha \leqq 1$ and

(i) $$p(z + \alpha v) \leqq pa(\omega) .$$

Obviously, $z + \alpha v \in S$. From the monotonicity of u we have

(ii) $$u(\omega, z + \alpha v) > u(\omega, z) .$$

By (i), we obtain $z + \alpha v \in R(\omega,p)$. Together with (ii) we conclude from this that

$$u(\omega,z) < \max_{x \in R(\omega,p)} u(\omega,x)$$

holds, which proves the lemma. □

2.2.5 Theorem: *Let* u *be monotonic. Then, every r-Walras allocation belongs to the r-core.*

Proof: Let (K,p) be an r-Walras-equilibrium . Further, let $L \in \Sigma$ and $A \in \mathcal{A}$ be given with

(i) $$\int_A \left(\int x \, L(\omega,dx) \right) d\mu \leqq \int_A a \, d\mu$$

and $L \succ_A K$. Since (K,p) is an r-Walras equilibrium, we have

(ii) $$K(\omega,\{z\in R(\omega,p) \mid u(\omega,z) = \max_{x\in R(\omega,p)} u(\omega,x)\}) = 1 \quad \mu\text{-a.e.}$$

From $L \succ_A K$ we obtain therefore that for μ-almost every $\omega \in A$ the relation

$$L(\omega,\{u(\omega,.) \geqq \alpha\}) \geqq K(\omega,\{u(\omega,.) \geqq \alpha\}) = 1$$

holds for every $\alpha \leqq \max_{x\in R(\omega,p)} u(\omega,x)$. Especially, we have

(iii) $$L(\omega,\{u(\omega,.) \geqq \max_{x\in R(\omega,p)} u(\omega,x)\}) = 1 \quad \mu\text{-a.e. within } A\,.$$

Observing Lemma 2.2.4 , this proves

(iv) $$L(\omega,\{z\in S \mid pz \geqq pa(\omega)\}) = 1 \quad \mu\text{-a.e. within } A\,.$$

If

(v) $$L(\omega,\{u(\omega,.) = \max_{x\in R(\omega,p)} u(\omega,x)\}) = 1 \quad \mu\text{-a.e. within } A$$

would hold, as a consequence of (ii), we would have

$$L(\omega,\{u(\omega,.) \geqq \alpha\}) = K(\omega,\{u(\omega,.) \geqq \alpha\}) \qquad (\alpha\in\mathbb{R})$$

μ-a.e. within A, which would imply $K \succsim_A L$ in contradiction to $L \succ_A K$. Hence, (v) cannot hold. Because of (iii), we can therefore find a set $B \in \mathcal{A}$ with $\mu(B) > 0$ and $B \subset A$ such that

$$L(\omega,\{u(\omega,.) > \max_{x\in R(\omega,p)} u(\omega,x)\}) > 0 \qquad (\omega\in B)$$

is satisfied. Consequently,

$$L(\omega,\{z\in S \mid pz > pa(\omega)\}) = L(\omega,S - R(\omega,p)) > 0 \qquad (\omega\in B)$$

holds. Together with (iv) we obtain from this

$$\int_A \left(\int pz \; L(\omega,dz) \right) d\mu > \int_A pa(\omega) \; d\mu$$

in contradiction to (i). This shows that no $L \in \Sigma$ exists with $L \succ_A K$ and (i). Therefore, K belongs to the r-core. □

This theorem proves especially in connection with 2.1.9 that the r-core is nonempty in case of monotonicity of u and balancedness of a. The problem of coincidence of the core and the set of Walras-allocations was already discussed by Vind (1964) under more general aspects. However, the methods used there seem to be not applicable to randomized allocations.

§ 3 SOME APPLICATIONS TO STATISTICAL DECISION THEORY

The mathematical models to be introduced in the sequel concern certain statistical decision problems. A statistician has carried out a random experiment and has to make a decision basing on the information gained by the experiment. The informations are controlled by a probability distribution which is not known to the statistician. Nevertheless, there is a collection of distributions under consideration, which come into question to control the random experiment. The space of possible results of the random experiment is assumed to be a measurable space $(\mathbb{H},\mathcal{H})$, which is called <u>sample space</u>. The collection of possible sample distributions is assumed to be given by a stochastic kernel

$$P : \Gamma\times\mathcal{H} \to [0,1] ,$$

where $(\Gamma,\mathcal{G})$ is a measurable space, and is called <u>parameter space</u>. Further, we presuppose that the collection

$$\mathcal{W} := \{P(\gamma,.)\,|\,\gamma\in\Gamma\}$$

of possible sample distributions is dominated by some probability measure μ on $\mathcal{H}$, and that there exists some measurable function

$$f : \mathbb{H}\times\Gamma \to \mathbb{R}_+$$

such that

$$P(\gamma,.) = f(.,\gamma)\,\mu \qquad (\gamma\in\Gamma) .$$

When the statistical decision problem is regarded as a game with incomplete information, it seems obvious that the probability space $(\mathbb{H},\mathcal{H},\mu)$ becomes the information space of the statistician. The density functions $f(.,\gamma)$ $(\gamma\in\Gamma)$ will be combined with the loss function which will be introduced later.

Nature is often regarded as counter-player of the statistician. This kind of naming should be well-understood in the sense that in most statistical decision problems the counter-player is rather an artificial person, who plays the role of a potential opponent of the statistician with decision space $(\Gamma,\mathcal{G})$. We assume that nature has no information space - formally one may introduce a one-point information space for this player.

The decision space of the statistician is denoted by $(S,\mathcal{S})$. This space can be of different feature for different kinds of decision problems. In estimation theory the space S is usually a subset of $\mathbb{R}^n$ containing certain characteristic values of the distributions $P(\gamma,.)$ $(\gamma\in\Gamma)$. In testing theory, S is equal to $\{0,1\}$, where 1 means the rejection of some hypothesis on the possible sample distribution, and 0 means non-rejection. In confidence estimation theory, S is a system of subsets of a given set Δ which often coincides with the parameter space Γ.

The strategy space of the statistician consists of all stochastic kernels $K : \mathbb{H}\times\mathcal{S} \to [0,1]$ and is denoted by Σ_E; similarly the strategy space of nature is the set of all probability measures on $\mathcal{G}$ and is denoted by Σ_N.

Finally, a <u>loss function</u>

$$v : \mathbb{H}\times S\times\Gamma \to \mathbb{R}$$

gives the loss of the statistician, when $x \in \mathbb{H}$ is the sample,

$s \in S$ is the decision of the statistician and $\gamma \in \Gamma$ is the unknown parameter value.

In the sequel, we base on the idea of Wald (1971) to consider the statistical decision problem, given by

$$(\mathbb{H},\mathcal{H},\mu;\, S,\mathcal{S};\, \Gamma,\mathcal{G};\, P,v) ,$$

as a game, and introduce a minimax-type solution concept. We will establish the connection to the theory of games with incomplete information as developped here.

3.1 Minimax decision rules

From the set-up of the statistical decision problem given here it presents itself to consider the problem as a game with incomplete information, and to transfer the solution concept of a Nash-equilibrium point to this problem. The corresponding solution concept for the statistical decision problem proves to consist in the minimax decision rule, which will be discussed in this section.

In order to apply the results concerning Nash-equilibrium points to the statistical decision problem, we assume that

(1) $(\mathbb{H}, H, \mu)$ is Polish and locally compact;

(2) $(S, \mathcal{S})$ and (Γ, G) are compact;

(3) the function $v^* : \mathbb{H} \times S \times \Gamma \to \mathbb{R}$ defined by $v^*(x,s,\gamma) := v(x,s,\gamma)\, f(x,\gamma)$ is a μ-C-function.

Condition (4) is for instance satisfied, when there are real numbers p and q with $1 \leq p$ and

$$\frac{1}{p} + \frac{1}{q} = 1$$

such that v^p and f^q are μ-C-functions, respectively.

For the described decision problem, the expected loss of the statistician is given by

$$V(K,Q) := \int v^* \, d(\mu \otimes K) \otimes Q \ ,$$

when strategies $K \in \Sigma_E$ and $Q \in \Sigma_N$ are used. The intention of the opponent nature is to confront the statistician with strategies as disadvantageous as possible for him. Thus, the game becomes a two-person zerosum game, and the expected payoff for the statistician is given by $U_E := -V$, that of nature is given by $U_N := V$. With this convention we obtain a game $\Gamma(U_E, U_N)$ as treated in Section 1.5 . This leads us to the following solution concept.

<u>3.1.1 Definition:</u> *Let $\varepsilon \geqq 0$ be given. Every strategy $\overline{K} \in \Sigma_E$ with*

$$\max_{Q \in \Sigma_N} V(\overline{K}, Q) - \varepsilon \leqq \max_{Q \in \Sigma_N} V(K, Q) \qquad (K \in \Sigma_E)$$

is called an <u>ε-minimax strategy</u> of the statistician. Every strategy $\overline{Q} \in \Sigma_N$ with

$$\min_{K \in \Sigma_E} V(K, Q) \leqq \min_{K \in \Sigma_E} V(K, \overline{Q}) + \varepsilon \qquad (Q \in \Sigma_N)$$

is called an <u>ε-least favourable prior distribution</u>.

In case $\varepsilon = 0$ we simply speak of a <u>minimax strategy</u> and a <u>least favourable prior distribution</u>, respectively.

Every pair $(\overline{K}, \overline{Q}) \in \Sigma_E \times \Sigma_N$ such that

$$\max_{Q \in \Sigma_N} V(\overline{K}, Q) - \varepsilon \leqq V(\overline{K}, \overline{Q}) \leqq \min_{K \in \Sigma_E} V(K, \overline{Q}) + \varepsilon$$

is called an <u>ε-saddle point</u> of the game and, respectively, a <u>saddle point</u> in case $\varepsilon = 0$.

<u>3.1.3 Remarks:</u> (1) The notation "max" and "min" instead of "sup" and "inf" in Definition 3.1.2 is justified by the fact that V is a continuous function on the set $\Sigma_E \times \Sigma_N$, which is a consequence of the assumptions and 1.5.4 .

(2) Let $\varepsilon \geq 0$ be given; and let $(\overline{K},\overline{Q})$ be an ε-saddle point. Then every $K \in \Sigma_E$ satisfies

$$\max_{Q\in\Sigma_N} V(\overline{K},Q) - \varepsilon \leq V(\overline{K},\overline{Q}) \leq V(K,\overline{Q}) + \varepsilon \leq$$

$$\leq \max_{Q\in\Sigma_N} V(K,Q) + \varepsilon ,$$

$\overline{K}$ proves therefore to be a 2ε-minimax strategy. Similarly, $\overline{Q}$ is a 2ε-least favourable prior distribution.

(3) Let $\varepsilon \geq 0$ be given; let $\overline{K}$ be an ε-minimax strategy and let $\overline{Q}$ be an ε-least favourable prior distribution. Assume further that an ε-saddle point exists. Then, a short estimation shows that $(\overline{K},\overline{Q})$ is a 4ε-saddle point, which is a certain kind of inversion of remark (2).

Since claims (1)-(3) imply that assumptions (1)-(6) in Section 1.5 are satisfied, we obtain from 1.5.7 an existence theorem for saddle points.

3.1.4 Theorem: *Conditions (1)-(3) imply the existence of a saddle point of the game* $\Gamma(U_E,U_N)$.

The purification problem can be solved for this game, whenever $\mathcal{W} = \{P(\gamma,.)|\gamma\in\Gamma\}$ is an atomless dominated class.

3.1.5 Theorem: *Let conditions (1)-(3) be satisfied, and let* μ *be nonatomic. Then, for every* $\varepsilon>0$ *there exists an* ε*-saddle point* $(\overline{K},\overline{Q})$*, where* $\overline{K}$ *is a pure strategy of the statistician.*

Proof: By 3.1.4 a saddle point $(K,\overline{Q})$ exists. The result is now obvious, since pure strategies are dense in Σ_E by 1.3.5, and since V is a uniformly continuous function on the compact set $\Sigma_E \times \Sigma_N$. □

Remark 3.1.3(2) shows that this is an existence theorem for a pure ε-minimax strategy,too. Condition (3) is somewhat stronger than the corresponding conditions mentioned in Eberl,Moeschlin (1982), 5.1.7, p. 161. Of course, 3.1.4 may be proved under weaker assumptions as claimed here, when a saddle point theorem of the type of Sion's Theorem (cp. for instance Theorem 7 in Aubin (1979), p. 218) is used. Nevertheless, 3.1.5 bases essentially on the uniform continuity of V, for which assumption (3) plays an important role. A result of the type of 3.1.5 was already given by Dvoretzky et.al. (1951), whose assumptions partially differ from those claimed here. Among other differences their loss function is not allowed to depend on the sample x.

In order to give a characterization of saddle points, we first introduce two notations.

3.1.6 Definition: *The function* $\rho : \Gamma \times \Sigma_E \to \mathbb{R}$ *defined by*

$$\rho(\gamma,K) := \int v^*(x,s,\gamma)\, d(\mu \otimes K)(x,s) \qquad (\gamma \in \Gamma)$$

is called the *prior risk.*

Analogously, the function $\varphi : \mathbb{H} \times S \times \Sigma_N \to \mathbb{R}$ *given by*

$$\varphi(x,s,Q) := \int v^*(x,s,\gamma)\, dQ(\gamma) \qquad (x \in \mathbb{H}, s \in S)$$

is called the *posteriori risk.*

Obviously the prior risk and posteriori risk are conditional payoffs for the statistician and his counter-player, respectively, as introduced in Section 1.5 . Lemma 1.5.8 yields the following characterization of saddle points.

<u>3.1.7 Lemma:</u> *For* $(K,Q) \in \Sigma_E \times \Sigma_N$ *the following two statements are equivalent:*

(1) (K,Q) *is a saddle point.*

(2) $\rho(.,K) = \max_{\gamma\in\Gamma} \rho(\gamma,K) \quad Q\text{-a.e.}$

and

$\varphi(x,.,Q) = \max_{s\in S} \varphi(x,s,Q) \quad K(x,.)\text{-a.e.} \quad \mu\text{-a.e.}$

We finally construct an example showing that even in case of an atomless dominated class the statistician has in general no pure minimax strategy. This example demonstrates that even under the special assumptions made in this section, we cannot expect the game to have an equilibrium point where one player uses a pure strategy. On the other hand, this justifies our approach as developped in the first five sections tending to approximate equilibrium points in pure strategies.

<u>3.1.8 Example:</u> Let the parameter space Γ be given by the union of two disjoint spheres

$$H_1 := \{m_1+z \mid z\in \mathbb{H}\} \text{ and } H_2 := \{m_2+z \mid z\in \mathbb{H}\}$$

where $\mathbb{H}$ is the unit sphere in $\mathbb{C}$; G designates the Borel σ-field over Γ. The sample space is $(\mathbb{H},H)$, where H is the Borel σ-field over $\mathbb{H}$. We denote the uniform distributions on $\mathbb{H},H_1$ and H_2 by

$\lambda_{\mathbb{H}}, \lambda_{H_1}$ and λ_{H_2}, respectively. Moreover, let c_1 and c_2 be two rational numbers with $1/\pi < c_1, c_2$. The functions $f_1, f_2 : \mathbb{H} \to \mathbb{R}_+$ are defined by

$$f_k(x) := \begin{cases} 2\pi c_k(1 - c_k|\arg x|) & \text{for } -\frac{1}{c_k} \leq \arg x \leq \frac{1}{c_k} \\ 0 & \text{elsewhere} \end{cases}$$

for $k = 1,2$, where arg x denotes the argument of x, when x is represented in polar coordinates. Thus we have

$$\int f_1 \, d\lambda_{\mathbb{H}} = \int f_2 \, d\lambda_{\mathbb{H}} = 1 .$$

We define the function $f : \mathbb{H} \times \Gamma \to \mathbb{R}_+$ by

$$f(x,\gamma) := \begin{cases} f_1(xz) & \text{for } \gamma = m_1 + z \in H_1 \\ f_2(xz) & \text{for } \gamma = m_2 + z \in H_2 \end{cases} \qquad (x \in \mathbb{H}) .$$

Since $\lambda_{\mathbb{H}}$ is invariant under rotations, we obtain a class of probability measures by setting

$$P(\gamma,.) = f(.,\gamma)\lambda_{\mathbb{H}} .$$

As decision space of the statistician we choose $S := \{1,2\}$ with the power set as σ-field. The loss function $v : S \times \Gamma \to \mathbb{R}$ is assumed to be defined by

$$v(s,\gamma) := \begin{cases} 0 & \text{for } s=1, \gamma \in H_1 \text{ and } s=2, \gamma \in H_2 \\ 1 & \text{for } s=1, \gamma \in H_2 \text{ and } s=2, \gamma \in H_1 . \end{cases}$$

The game becomes therefore a testing problem with H_1 as hypothesis and H_2 as alternative.

Conditions (1)-(3) are satisfied for this example. Hence, by 3.1.4, the game has a saddle point. By an application of

3.1.7 we check that (K,Q) with

$$K(x,s) := K(x,\{s\}) := \frac{1}{2} \quad (x\in\mathbb{H},\ s\in\{1,2\})$$

and

$$Q := \frac{1}{2}\lambda_{H_1} + \frac{1}{2}\lambda_{H_2}$$

is a saddle point of the game. Moreover, we have:

There is no pure minimax strategy of the statistician.

Proof: Let ε_g be a pure strategy of the statistician; and let

$$A_s := \{x\in\mathbb{H} \mid g(x)=s\} \quad (s=1,2)\ .$$

Then, $A_1 + A_2 = \mathbb{H}$ holds, and

$$\text{(i)} \qquad \rho(\gamma,\varepsilon_g) = \int v(g(x),\gamma)\ f(x,\gamma)\ d\lambda_{\mathbb{H}}(x) = \begin{cases} \int_{A_2} f(x,\gamma)\ d\lambda_{\mathbb{H}}(x) & \text{for } \gamma\in H_1\ , \\ \int_{A_1} f(x,\gamma)\ d\lambda_{\mathbb{H}}(x) & \text{for } \gamma\in H_2\ . \end{cases}$$

Suppose, ε_g is a minimax strategy. Then, by 3.1.3(2) and 3.1.3(3), (ε_g,Q) is a saddle point. In particular, $\rho(.,\varepsilon_g)$ has to be constant Q-a.e. by 3.1.7 . We will show that this cannot be true. Otherwise, since

$$\int_{A_2} f(x,\gamma)\ d\lambda_{\mathbb{H}}(x) = \int_{A_2} f_1(xz)\ d\lambda_{\mathbb{H}}(x) =$$

$$= \int 1_{A_2}(x)\ f_1(xz)\ d\lambda_{\mathbb{H}}(x) \qquad (\gamma = m_1+z \in H_1)\ ,$$

we derive from (i) that the last integral takes the same value for $\lambda_{\mathbb{H}}$-almost every $z\in\mathbb{H}$. Define $\mu := 1_{A_2}\lambda_{\mathbb{H}}$. By a continuity argument we even have

$$\int f_1(xz)\, d\mu(x) = \text{const} \qquad (z \in \mathbb{H}) .$$

As f_1 is lefthanded differentiable (cp. A.4.1) and - as it is easy to check - the assumptions of A.4.2 are satisfied,

$$\text{(ii)} \qquad \int D_1 f_1(xz)\, d\mu(x) = 0 \qquad (z \in \mathbb{H})$$

holds. Setting

$$B_\varphi(z) := \{ze^{i\delta} \mid 0 \leq \delta < \varphi\} \qquad (\varphi \in \mathbb{R}_+ ,\ z \in \mathbb{H}) ,$$

we verify by definition of f_1 that

$$D_1 f_1(xz) = 1_{B_{\frac{1}{c_1}}(ze^{-1/c_1})}(x) - 1_{B_{\frac{1}{c_1}}(z)}(x) \qquad (x, z \in \mathbb{H})$$

holds. Together with (ii) this implies

$$\text{(iii)} \qquad \mu(B_{\frac{1}{c_1}}(z)) = \mu(B_{\frac{1}{c_1}}(ze^{i/c_1})) \qquad (z \in \mathbb{H}) .$$

By rationality of c_1, the number $1/2\pi c_1$ is irrational. From (iii) and A.4.3 we conclude that μ is proportional to $\lambda_{\mathbb{H}}$. Hence, $\lambda_{\mathbb{H}}(A_2) = 0$ or $\lambda_{\mathbb{H}}(A_2) = 1$. In the first case we obtain by (i)

$$\rho(\gamma, \varepsilon_g) = \begin{cases} 0 & \text{for } \gamma \in H_1 \\ 1 & \text{for } \gamma \in H_2 \end{cases}$$

contradicting the fact that $\rho(\cdot, \varepsilon_g)$ is Q-a.e. constant. The second case results in the same contradiction. □

3.2 Set-valued minimax estimators

We apply now the results of Section 3.1 to a decision problem for set-valued decision functions. Our approach differs essentially from the standard concepts for set-valued estimators as exposed for instance in Berger (1980) or Rohatgi (1976), i.e. the concepts of credible regions and confidence intervals, respectively. We discuss two different minimax decision problems for set-valued estimators basing on two different types of loss functions. The first approach is inspired by a certain kind of testing procedure and is related to a concept of Bayesian set-valued estimators as developped by Moeschlin (1982). The second approach bases on the Hausdorff-topology on the system of convex and compact subsets of a compact metric space. For both decision problems pure strategies play an outstanding role, because they can be identified with set-valued estimators.

The estimation problem to be considered in the sequel has the special quality that the decision space of the statistician is a nonempty system $\mathcal{D}$ of subsets of a nonempty set Δ (in many situations Δ coincides with the parameter set Γ).

In our first approach we assume $\mathcal{D}$ to be a σ-field over Δ. Moreover, we assume to be given a probability measure ν on $\mathcal{D}$, i.e. we start with a probability space $(\Delta,\mathcal{D},\nu)$. The loss function to be introduced for this specific decision problem bases on the idea that the statistician decides for each $\delta \in \Delta$ whether it belongs to the set he chooses or not. His decision $D \in \mathcal{D}$ consists therefore of all $\delta \in \Delta$ which have been accepted. The loss appearing during this decision process is given by a function

$$v : \Delta \times \Gamma \to [0,1]$$

which has the following interpretation. The value $v(\delta,\gamma)$ designates the loss, when $\gamma \in \Gamma$ is the underlying parameter and $\delta \in \Delta$ is accepted, while $1 - v(\delta,\gamma)$ is the loss, when $\delta \in \Delta$ is rejected.

Every set-valued mapping $Z : \mathbb{H} \to \mathcal{D}$ with an $\mathcal{H}\otimes\mathcal{D}$-measurable graph

$$G_Z := \{(x,\delta) \in \mathbb{H}\times\Delta \mid \delta \in Z(x)\}$$

is called a <u>set-valued estimator</u> (SVE) in this model. The set of all strategies of the counter-player is the set Σ_N of all probability measures on $\mathcal{G}$ as in Section 3.1 . The expected loss of the statistician, when the SVE Z and the strategy $Q \in \Sigma_N$ are used, is given by

$$\begin{aligned} V^*(Z,Q) &:= \\ &= \int \Big(\int \Big(\int_{Z(x)} v(\delta,\gamma)\, d\nu(\delta) + \int_{(Z(x))^c} (1-v(\delta,\gamma))\, d\nu(\delta) \Big) \\ &\qquad f(x,\gamma)\, d\mu(x) \Big)\, dQ(\gamma) = \\ &= \int \Big(\int_{G_Z} v(\delta,\gamma)\, f(x,\gamma)\, d\mu\otimes\nu(x,\delta) + \int_{G_Z^c} (1-v(\delta,\gamma))\, d\mu\otimes\nu(x,\delta) \Big)\, dQ(\gamma)\ . \end{aligned}$$

Roughly spoken, V^* measures the bad values $\delta \in \Delta$ in $Z(x)$, i.e. those with high loss, together with the good values in $(Z(x))^c$, i.e. those with low loss.

We introduce a minimax solution concept for the given estimation problem.

<u>3.2.1 Definition:</u> *Let Ψ be the set of all SVEs. For given $\varepsilon \geqq 0$ the SVE $\overline{Z} \in \Psi$ is called an <u>ε-minimax SVE</u>, if*

$$\max_{Q \in \Sigma_N} V^*(\overline{Z},Q) - \varepsilon \leq \max_{Q \in \Sigma_N} V^*(Z,Q) \qquad (Z \in \Psi)$$

holds. In case $\varepsilon=0$, $\overline{Z}$ is called a <u>minimax SVE</u>.

We consider the statistical decision problem $(\mathbb{H}',H',\mu\otimes\nu,\{0,1\}, P\{0,1\};\Gamma,G;P',v')$ with $(\mathbb{H}',H') := (\mathbb{H}\times\Delta,H\otimes\mathcal{D})$, $P'(\gamma,.) := P(\gamma,.)\otimes\nu \quad (\gamma\in\Gamma)$ and

$$v'(\delta,s,\gamma) := \begin{cases} v(\delta,\gamma) & \text{for } s=0 \\ 1 - v(\delta,\gamma) & \text{for } s=1 . \end{cases}$$

The set of all strategies of the statistician is designated by Σ_E as in the previous section. The expected loss of the statistician is then given by

$$V(K,Q) :=$$
$$= \int (\int v'(\delta,s,\gamma)\ f(x,\gamma)\ K(x,\delta;ds))\ d\mu\otimes\nu\otimes Q \qquad (K\in\Sigma_E,\ Q\in\Sigma_N) .$$

The SVEs $Z \in \Psi$ can now be identified with pure strategies $K \in \Sigma_E$ by virtue of

$$\delta \in Z(x) \iff K(x,\delta;\{0\}) = 1$$

for every $(x,\delta) \in \mathbb{H}\times\Delta$. We denote the pure strategy associated with Z by K_Z, and, conversely, the SVE Z associated with the pure strategy $K \in \Sigma_E$ is denoted by Z_K. Thus we have

$$V^*(Z,Q) = V(K_Z,Q) \qquad (Q\in\Sigma_N) .$$

We assume in the sequel that

(1)' $(\mathbb{H}\times\Delta, H\otimes\mathcal{D}, \mu\otimes\nu)$ is a Polish and locally compact information space;

(2)' (Γ, G) is compact;

(3)' the function $\bar{v}: \mathbb{H}\times\Delta\times S\times\Gamma \to \mathbb{R}$ given by $\bar{v}(x,\delta,s,\gamma) := v'(\delta,s,\gamma)f(x,\gamma)$ is a $\mu\otimes\nu$-C-function.

Assumptions (1)' - (3)' ensure that assumptions (1)-(3) in 3.1 are satisfied for the statistical decision problem discussed here.

Since pure strategies and SVEs can be identified, we first verify the following fact.

3.2.2 Lemma: *For given* $\varepsilon \geq 0$ *and every pure* ε*-minimax strategy* $K \in \Sigma_E$ *the SVE* Z_K *is an* ε*-minimax SVE.*

Proof: The statement follows from the fact

$$V^*(Z_{K'}, Q') = V(K', Q') \qquad (Q' \in \Sigma_E)$$

for all pure strategies $K' \in \Sigma_E$ and the property

$$\max_{Q'\in\Sigma_N} V(K,Q') - \varepsilon \leq \max_{Q'\in\Sigma_N} V(K',Q')$$

for every pure strategy $K' \in \Sigma_E$. □

Together with Theorem 3.1.5 we obtain now an existence result for ε-minimax SVEs.

3.2.3 Theorem: *Let assumptions (1)' - (3)' be satisfied and let the measure* $\mu\otimes\nu$ *be nonatomic. Then for* $\varepsilon > 0$ *an* ε*-minimax SVE exists.*

The proof is obvious.

We finally give an example, where even a minimax SVE exists.

3.2.4 Example: We assume all spaces $\mathbb{H}, \Delta$ and Γ to be compact subsets of $\mathbb{R}^n$ and the function $f : \mathbb{H}\times\Gamma \to \mathbb{R}$ to be a polynomial in all appearing variables - this assumptions has its justification in the fact that densities can often be approximated by polynomials in a suitable way. Further, we assume that the loss $v : \Delta\times\Gamma \to [0,1]$ is also a polynomial in the appearing components. Assuming this, we immediately realize that the function $\overline{v}$ defined by

$$\overline{v}(x,\delta,s,\gamma) = \begin{cases} v(\delta,\gamma)\ f(x,\gamma) & \text{for } s=0 \\ (1 - v(\delta,\gamma))f(x,\gamma) & \text{for } s=1 \end{cases}$$

becomes a sum of tensor products. Therefore, whenever $\mu\otimes\nu$ is a nonatomic measure, the same method as used to prove Lemma 1.6.3 shows that every minimax strategy K has a purification K_Z, i.e.

$$V(K_Z,Q) = V(K,Q) \qquad (Q\in\Sigma_N)$$

holds. Hence, Z is a minimax SVE.

An alternative approach to set-valued estimation problems is based on the idea to consider the system $\mathcal{D}$ of subsets of Δ itself as decision space of the statistician. To this end we introduce some structure on $\mathcal{D}$ which allows to define a suitable σ-field on $\mathcal{D}$. Such a structure exists, when Δ is a compact metric space and $\mathcal{D}$ equals the system $C(\Delta)$ of nonempty compact subsets of Δ. Thus, $C(\Delta)$ becomes a compact metric space w.r.t. the corresponding Hausdorff-distance (cp.A.2.1). If $\mathcal{C}(\Delta)$ is endowed with the corresponding Borel σ-field $\mathcal{B}(\mathcal{C}(\Delta))$, then $(S,\mathcal{S}) := (\mathcal{C}(\Delta),\mathcal{B}(\mathcal{C}(\Delta)))$ becomes a compact decision space for the statistician. A set-valued estimator (SVE) for this model is an $\mathcal{H}$-$\mathcal{B}(\mathcal{C}(\Delta))$-measurable mapping $Z : \mathbb{H} \to \mathcal{C}(\Delta)$.

We assume now that

(1)" $(\mathbb{H},\mathcal{H},\mu)$ is a Polish and locally compact information space

and that

(2)" $(\Gamma,\mathcal{G})$ is compact.

Further, we assume a continuous correspondence $L : \Gamma \to \mathcal{C}(\Delta)$, which has the following interpretation: $L(\gamma)$ is the set of those $\delta \in \Delta$ which are acceptable for the statistician when $\gamma \in \Gamma$ is the true parameter value. For a given continuous function $w : \mathcal{C}(\Delta)\times\mathcal{C}(\Delta) \to \mathbb{R}$ we obtain a continuous loss function $v_w : \mathcal{C}(\Delta)\times\Gamma \to \mathbb{R}$ by

$$v_w(C,\gamma) := w(C,L(\gamma)) \quad (C\in\mathcal{C}(\Delta),\gamma\in\Gamma).$$

Assuming in addition that

(3)" f is a μ-C-function,

we verify that assumptions (1)-(3) in 3.1 are satisfied for the given statistical decision problem. Therefore, conditions (1)" - (3)" are sufficient for the existence of an ε-minimax SVE defined according to 3.2.1.

It deserves to be remarked that this kind of set-estimation problem often allows a reduction to a point-estimation problem.

For instance, if we choose especially $\Delta := \Gamma := [0,1]$ and

$$L(\gamma) := \{\delta \in [0,1] \mid \delta \leqq \gamma\} \quad (\gamma \in [0,1)\,,$$

and if we introduce the loss function by the Hausdorff-distance d on $\mathcal{C}(\Delta)$, we obtain

$$d(L(\gamma), L(\gamma')) = |\gamma - \gamma'| \quad (\gamma, \gamma' \in \Gamma)\,,$$

and, thus we are led to a common point estimation problem with linear loss. It can even be shown that every pure ε-minimax strategy for the point estimation problem gives a corresponding ε-minimax SVE.

APPENDIX

In the sequel some results from different mathematical fields will be summarized in order to make the treatise self-contained. Proofs will be largely omitted for the reason of size. However, there will be references to the corresponding literature. The results are arranged according to the fields they are attached to, and are partially based on each other.

A1 Measure-theoretic concepts

First, we assume a metric space E together with the corresponding Borel σ-field $\mathcal{B}(E)$ over E. Further, $\mathcal{W}(E)$ designates the set of all probability measures on $\mathcal{B}(E)$. On the set $\mathcal{W}(E)$ we assume the topology of weak convergence. For $P_n \in \mathcal{W}(E)$ $(n \in \mathbb{N} \cup \{0\})$ we have thereby

$$\lim_{n \to \infty} P_n = P_0$$

if and only if

$$\lim_{n \to \infty} \int h \, dP_n = \int h \, dP_0$$

holds for every bounded continuous function $h : E \to \mathbb{R}$. The following theorem, known as Portmanteau Theorem, can be found in Billingsley (1968) as Theorem 2.1 on page 11.

A.1.1 Theorem: *Let* $P_n \in \mathcal{W}(E)$ $(n \in \mathbb{N} \cup \{0\})$ *be given. Then, the following four assertions are equivalent:*

(1) $\lim_{n \to \infty} P_n = P_0$;

(2) $\lim_{n\to\infty} \int h \, dP_n = \int h \, dP_o$ *for every bounded uniformly continuous function* $h : E \to \mathbb{R}$;

(3) $\overline{\lim} \, P_n(A) \leq P_o(A)$ *for every closed set* $A \subset E$;

(4) $\underline{\lim} \, P_n(U) \geq P_o(U)$ *for every open set* $U \subset E$.

For separable metric spaces, assertion (2) can be modified in the following way:

<u>A.1.2 Lemma:</u> *Let* E *be a totally bounded sparable metric space. Then, a sequence* $(h_k | k \in \mathbb{N})$ *of bounded, uniformly continuous functions* $h_k : E \to \mathbb{R}$ *exist with the property:*

for every sequence $(P_n | n \in \mathbb{N} \cup \{0\})$ *from* $W(E)$ *the assertions*

(1) $\lim_{n\to\infty} P_n = P_o$

and

(2) $\lim_{n\to\infty} \int h_k \, dP_n = \int h_k \, dP_o \quad (k \in \mathbb{N})$

are equivalent.

(cp. Parthasarathy (1967), II. Th. 6.6, p.47).

It should be emphasized that by a theorem of Urysohn (cp. Kelly (1961), p. 125) every separable metric space allows an equivalent totally bounded metrization.

The sequence $(h_k | k \in \mathbb{N})$ appearing in A.1.2 can additionally be assumed to be uniformly bounded. Otherwise this can be ensured, when the sequence (h_k) is substituted by (h_k') with

$$h'_k := \begin{cases} \dfrac{1}{\sup\limits_{x\in E}|h_k(x)|}\, h_k & \text{for } \sup\limits_{x\in E}|h_k(x)| > 1 \\ h_k & \text{else .} \end{cases}$$

Thus, we obtain the following corollary of A.1.2 (cp. also Nowak (1985), Lemma 4.1).

A.1.3 Corollary: *Let* E *be a totally bounded separable metric space. Then, a sequence* $(h_k|k\in\mathbb{N})$ *of uniformly continuous functions* $h_k : E \to [-1,1]$ *exists such that the metric* ρ *defined by*

$$\rho(P,Q) := \sum_{k=1}^{\infty} \frac{1}{2^k} \left| \int h_k \, dP - \int h_k \, dQ \right|$$

generates the weak topology on $\mathcal{W}(E)$.

The proof is obvious.

The following result is a consequence of A.1.3 and II. Theorem 6.2, p. 43, in Parthasarathy (1967).

A.1.4 Theorem: *Let* E *be a separable metric space. Then,* $\mathcal{W}(E)$ *endowed with the metric* ρ *as defined in A.1.3, is a separable metric space.*

The relatively compact subsets of $\mathcal{W}(E)$ may be characterized by means of II. Theorem 6.7, p. 67, in Parthasarathy (1967).

A.1.5 Theorem: *Let* E *be a separable metric space; and let* $\mathcal{V}$ *be a subset of* $\mathcal{W}(E)$. *Then, the following two statements imply each other:*

(1) $\mathcal{V}$ *is relatively compact.*

(2) *For every* $\varepsilon>0$, *there exists a compact set* $C_\varepsilon \subset E$ *such that* $P(C_\varepsilon) \geq 1-\varepsilon$ *holds for every* $P \in V$.

Theorem A.1.5 is known as Theorem of Prohorov (cp. also Billingsley (1968), Theorem 6.1, 6.2 on page 37).

The following lemma results from Theorem 3.2, p. 21, in Billingsley (1968) by induction on the number of factors.

<u>A.1.6 Lemma:</u> *Let* $E_1,\ldots,E_N$ *be separable metric spaces. Then, the mapping defined by*

$$(P_1,\ldots,P_N) \longmapsto P_1\otimes\ldots\otimes P_N$$

from $W(E_1)\times\ldots\times W(E_N)$ *into the space* $W(E_1\times\ldots\times E_N)$ *is continuous w.r.t. the product of the weak topologies on* $W(E_1),\ldots,W(E_N)$ *and the weak topology on* $W(E_1\times\ldots E_N)$.

The next four theorems concern some measurability results. The first theorem is a variant of Lusin's Theorem, as to be found for instance in Bauer (1977) as Satz 41.4 on page 202.

<u>A.1.7 Lusin's Theorem:</u> *Let* E *be a Polish space, let* E' *be a topological space with a countable base, and let* $P \in W(E)$. *Then, for every mapping* $f: E \to E'$ *the following two properties are equivalent:*

(1) f *coincides* P-*a.e. with some* $\mathcal{B}(E)$-$\mathcal{B}(E')$-*measurable mapping defined on* E *with values in* E'.

(2) *For every* $\varepsilon>0$ *a compact set* $C \subset E$ *with* $P(C) \geq 1-\varepsilon$ *exists such that* f *is continuous on* C.

Henceforth, we assume a probability space (Ω, A, P) and a separable metric space S endowed with its Borel σ-field $\mathcal{B}(S)$.

The idea of proof for the following assertion is due to Kuratowski (1966), p. 378 (cp. also Himmelberg (1975), Theorem 6.1).

<u>A.1.8 Lemma:</u> *Let* $u : \Omega \times S \to \mathbb{R}$ *be a function for which*

$$u(\omega,.) \text{ is continuous} \quad (\omega \in \Omega)$$

and

$$u(.,s) \text{ is } A\text{-}\mathcal{B}(\mathbb{R})\text{-measurable} \quad (s \in S) .$$

Then, u *is* $A \otimes \mathcal{B}(S) - \mathcal{B}(\mathbb{R})$*-measurable* .

Proof: Let $A \subset \mathbb{R}$ be closed, and let D be a countable dense subset of S. Further, let d be the Euclidean metric on $\mathbb{R}$, and let d_S be the metric on S. Since $u(\omega,.)$ is continuous, we have $u(\omega,s) \in A$ if and only if for every $n \in \mathbb{N}$ there exists some $t \in D$ with $d_S(s,t) \leq \frac{1}{n}$ and $d(u(\omega,t),A) \leq \frac{1}{n}$. Therefore

$$u^{-1}(A) = \bigcap_{n \in \mathbb{N}} \bigcup_{t \in D} \{\omega \in \Omega \mid d(u(\omega,t),A) \leq \tfrac{1}{n}\} \times$$
$$\times \{s \in S \mid d_S(s,t) \leq \tfrac{1}{n}\} \in A \otimes \mathcal{B}(S) .$$

This shows that u is an $A \otimes \mathcal{B}(S)$-$\mathcal{B}(\mathbb{R})$ measurable function. □

The following theorem is a generalization of Lusin's result in some sense. A proof was given by Castaing (1967), Theorem 3.1 .

<u>A.1.9 Theorem:</u> *Let* Ω *and* S *be locally compact and Polish spaces. Further, let* $A = \mathcal{B}(\Omega)$ *and let* $u : \Omega \times S \to \mathbb{R}$ *be a function such that*

$$u(\omega,.) \text{ is continuous} \quad (\omega \in \Omega)$$

and

$$u(.,s) \text{ is } \mathcal{A}\text{-}\mathcal{B}(\mathbb{R})\text{-measurable} \quad (s \in S) .$$

Then, for every $\varepsilon > 0$ *a compact set* $C \subset \Omega$ *with* $P(C) \geq 1 - \varepsilon$ *exists such that* u *restricted to* $C \times S$ *is continuous.*

This slightly stronger result in comparison with that given by Castaing follows easily from the additional claim that Ω has to be Polish which implies the inner regularity of the measure P on $\mathcal{B}(\Omega)$. If S is a one-point set, A.1.9 becomes a version of Lusin's Theorem.

Further, the following fact deserves to be mentioned.

<u>A.1.10 Lemma:</u> *Let* $u : \Omega \times S \to \mathbb{R}$ *be an* $\mathcal{A} \otimes \mathcal{B}(S) - \mathcal{B}(\mathbb{R})$*-measurable function, and let* $F : \Omega \to \mathcal{B}(S)$ *be a correspondence with*

$$\text{Graph}(F) \in \mathcal{A} \otimes \mathcal{B}(S) .$$

Further, let $\hat{\mathcal{A}}$ *be the completion of* $\mathcal{A}$ *w.r.t.* P. *Then, the function* $\overline{u}$ *given by*

$$\overline{u}(\omega) := \sup_{s \in F(\omega)} u(\omega, s) \quad (\omega \in \Omega)$$

with values in the extended real numbers $\overline{\mathbb{R}}$ *is* $\hat{\mathcal{A}}\text{-}\mathcal{B}(\overline{\mathbb{R}})$*-measurable.*

Proof: For every $\alpha \in \mathbb{R}$, we have

$$\{\overline{u} > \alpha\} = \{\omega \in \Omega \mid \exists\, s \in F(\omega) : u(\omega, s) > \alpha\} =$$

$$= \pi(\text{Graph}(F) \cap \{u > \alpha\}) ,$$

where π is the projection of $\Omega \times S$ onto S. From the Projection Theorem we conclude that

$$\{\overline{u} > \alpha\} \in \hat{\mathcal{A}}$$

holds for every $\alpha \in \mathbb{R}$. This proves $\hat{\mathcal{A}}\text{-}\mathcal{B}(\overline{\mathbb{R}})$-measurability of $\overline{u}$.□

Endowing the space $\mathcal{W}(S)$ with the Borel σ-field $\mathcal{B}(\mathcal{W}(S))$ generated by the system of open subsets of $\mathcal{W}(S)$, we obtain the following result.

<u>A.1.11 Lemma:</u> *For every mapping* $K : \Omega \to \mathcal{W}(S)$ *the following both statements are equivalent:*

(1) K *is* $\mathcal{A}$-$\mathcal{B}(\mathcal{W}(S))$*-measurable.*

(2) *The mapping* $K' : \Omega \times \mathcal{B}(S) \to [0,1]$ *given by*

$$K'(\omega,A) := (K(\omega))(A) \quad (\omega \in \Omega, A \in \mathcal{B}(S))$$

is a stochastic kernel.

Proof: By Rieder (1975), 6.1, the mapping $K : \Omega \to \mathcal{W}(S)$ is $\mathcal{A}$-$\mathcal{B}(\mathcal{W}(S))$-measurable, if and only if each of the mappings

$$\omega \to (K(\omega))(A) \quad (A \in \mathcal{B}(S))$$

is $\mathcal{A}$-$\mathcal{B}(\mathbb{R})$-measurable. This is in turn satisfied, if and only if K' is a stochastic kernel. □

The following lemma is a generalization of Bienaymé's equation (cp. Bauer (1977), Satz 32.3, p. 154).

<u>A.1.12 Lemma:</u> *Let the sequence* $(X_n \mid n \in \mathbb{N})$ *consist of pairwise uncorrelated random variables* $X_n : (\Omega, \mathcal{A}, P) \to (\mathbb{R}, \mathcal{B})$ *such that*

$$\sum_{n \in \mathbb{N}} |X_n| \text{ is quadratically } P\text{-integrable.}$$

Then, the equation

$$V\Big(\sum_{n \in \mathbb{N}} X_n\Big) = \sum_{n \in \mathbb{N}} V(X_n)$$

exists.

Proof: As a consequence of the dominated convergence theorem, $\sum_{n \in \mathbb{N}} X_n$ is P-integrable and

(i) $$E(\sum_{n \in \mathbb{N}} X_n) = \sum_{n \in \mathbb{N}} E(X_n) .$$

The same argument shows that $\sum_{n \in \mathbb{N}} X_n$ is even quadratically P-integrable and that

(ii) $$E((\sum_{n \in \mathbb{N}} X_n)^2) = \lim_{m \to \infty} E((\sum_{n=1}^{m} X_n)^2)$$

holds. From (i) and (ii), we obtain by Bienaymé's equation for finitely many uncorrelated random variables that the sequence of equations

$$V(\sum_{n \in \mathbb{N}} X_n) = E((\sum_{n \in \mathbb{N}} X_n)^2) - (E(\sum_{n \in \mathbb{N}} X_n))^2 =$$

$$= \lim_{m \to \infty} E((\sum_{n=1}^{m} X_n)^2) - \lim_{m \to \infty} (\sum_{n=1}^{m} E(X_n))^2 =$$

$$= \lim_{m \to \infty} V(\sum_{n=1}^{m} X_n) = \lim_{m \to \infty} \sum_{n=1}^{m} V(X_n) =$$

$$= \sum_{n \in \mathbb{N}} V(X_n)$$

holds. □

A 2 Measurable correspondences

Let be given a metric space E. Further, let S be a compact metric space w.r.t. some metric d, and let $\mathcal{B}(S)$ be the corresponding Borel σ-field. The system of all compact nonvoid subsets of S is denoted by $\mathcal{C}(S)$. We assume the Hausdorff distance d^H on $\mathcal{C}(S)$, which is defined by

$$d^H(A,B) := \max(\max_{a\in A} d(a,B), \max_{b\in B} d(b,A)) \quad (A,B \in \mathcal{C}(S)).$$

The following theorem is due to Hausdorff (1957), prop.VI, p. 172.

<u>A.2.1 Theorem:</u> *$\mathcal{C}(S)$ is a compact metric space w.r.t. the Hausdorff distance.*

If we introduce the Hausdorff semidistance d_o^H by

$$d_o^H(A,B) := \max_{a\in A} d(a,B) \quad (A,B \in \mathcal{C}(S)),$$

we immediately realize that

$$(*) \qquad d^H(A,B) = \max(d_o^H(A,B), d_o^H(B,A)) \quad (A,B \in \mathcal{C}(S)).$$

<u>A.2.2 Definition:</u> *Let $F: E \to \mathcal{C}(S)$ be a correspondence. Further, let $t_o \in E$. The correspondence F is said to be <u>upper semicontinuous</u> (<u>lower semicontinuous</u> and <u>continuous</u>, resp.) at t_o, iff the relation*

$$\lim_{n\to\infty} d_o^H(F(t_n), F(t_o)) = 0$$

$$(\lim_{n\to\infty} d_o^H(F(t_o), F(t_n)) = 0 \text{ and } \lim_{n\to\infty} d^H(F(t_n), F(t_o)) = 0, \text{ resp.})$$

holds for every sequence (t_n) *from* E *converging to* t_o.

The correspondence F *is said to be* <u>*upper semicontinuous*</u> *(*<u>*lower semicontinuous*</u> *and* <u>*continuous*</u>*, resp.), iff* F *is upper semicontinuous (lower semicontinuous and continuous, resp.) at every point* $t_o \in E$.

As a consequence of (*), a correspondence $F : E \to C(S)$ is continuous, if and only if it is upper and lower semicontinuous. Upper and lower semicontinuity of a correspondence F can be characterized in the following way:

<u>A.2.3 Lemma:</u> *Let* $F : E \to C(S)$ *be a correspondence, and let* $t_o \in E$. *Then, the following assertions are equivalent:*

(1) F *is upper semicotinuous at* t_o.

(2) *For every sequence* (t_n) *from* E *converging to* t_o *and every convergent sequence* (s_n) *from* S *with* $s_n \in F(t_n)$ $(n \in \mathbb{N})$, *the relation* $\lim_{n\to\infty} s_n \in F(t_o)$ *holds*.

Moreover, the following two assertions are equivalent:

(1)' F *is lower semicontinuous at* t_o.

(2)' *Let* $s_o \in F(t_o)$, *and let* U *be some neighbourhood of* s_o. *Then, for every sequence* (t_n) *from* E *converging to* t_o, *the relation* $F(t_n) \cap U \neq \emptyset$ *holds for almost every* $n \in \mathbb{N}$.

Proof: (1) ⟹ (2): Let F be upper semicontinuous at t_o. Further, let $\lim_{n\to\infty} t_n = t_o$, $s_n \in F(t_n)$ $(n \in \mathbb{N})$ and let $s_o = \lim_{n\to\infty} s_n$. Then, because of

$$d(s_n, F(t_o)) \leq \max_{s \in F(t_n)} d(s, F(t_o)) = d_o^H(F(t_n), F(t_o)),$$

we obtain

$$d(s_o, F(t_o)) = \lim_{n\to\infty} d(s_n, F(t_o)) = 0 .$$

Since $F(t_o)$ is compact, this proves $s_o \in F(t_o)$, hence (2).

(2) ⟹ (1): Let $\lim_{n\to\infty} t_n = t_o$; and let $s_n \in F(t_n)$ s.t.

$$\text{(i)} \qquad d(s_n, F(t_o)) = \max_{s\in F(t_n)} d(s, F(t_o))$$

holds. Then, a subsequence (s_{n_k}) of (s_n) exists with

$$\lim_{k\to\infty} d(s_{n_k}, F(t_o)) = \overline{\lim}\, d(s_n, F(t_o)) .$$

W.l.o.g., since S is compact, we may assume that (s_{n_k}) converges to some $s_o \in S$. From (2), we derive $s_o \in F(t_o)$. Therefore, we have

$$\overline{\lim}\, d(s_n, F(t_o)) = \lim_{k\to\infty} d(s_{n_k}, F(t_o)) = d(s_o, F(t_o)) = 0 .$$

Together with (i), this proves the validity of (1).

The second equivalence statement follows by similar methods. □

The following theorem is borrowed from Hildenbrand (1974), p. 30.

<u>A.2.4 Theorem:</u> *Let* $F : E \to C(S)$ *be a continuous correspondence, and let* $u : E\times S \to \mathbb{R}$ *be a continuous function. Then, the function* $\overline{u} : E \to \mathbb{R}$ *given by*

$$\overline{u}(t) := \max_{s\in F(t)} u(t,s) \qquad (t\in E)$$

is continuous, and the correspondence $G : E \to C(S)$ *given by*

$$G(t) := \{s\in F(t) \mid u(t,s) = \overline{u}(t)\} \qquad (t\in E)$$

is upper semicontinuous.

Proof: Let $t_o \in E$, and let (t_n) be a sequence from E with $\lim\limits_{n\to\infty} t_n = t_o$. For every $n \in \mathbb{N}$ let $s_n \in F(t_n)$ be given such that

$$u(t_n,s_n) = \overline{u}(t_n) .$$

Take a subsequence (s_{n_k}) of (s_n) with the property

$$\lim_{k\to\infty} u(t_{n_k},s_{n_k}) = \overline{\lim}\, \overline{u}(t_n) .$$

Since S is compact, we may even assume that there is some $s_o = \lim\limits_{k\to\infty} s_{n_k} \in S$. From the continuity of u we obtain

$$\text{(i)} \qquad u(t_o,s_o) = \overline{\lim}\, \overline{u}(t_n) .$$

Since F is upper semicontinuous, an application of A.2.3 ((1) $\Longleftrightarrow$ (2)) shows $s_o \in F(t_o)$. Together with (i), this implies

$$\text{(ii)} \qquad \overline{u}(t_o) \geq u(t_o,s_o) = \overline{\lim}\, \overline{u}(t_n) .$$

Let now $\varepsilon > 0$ be given; and let $s_1 \in F(t_o)$ be such that

$$u(t_o,s_1) = \overline{u}(t_o) .$$

Since u is continuous, we find a neighbourhood U of s_1 and a neighbourhood V of t_o with the property

$$|u(t,s) - u(t_o,s_1)| \leq \varepsilon \quad (t\in V,\ s\in U) .$$

As a consequence of the lower semicontinuity of F and A.2.3 ((1)' $\Longleftrightarrow$ (2)'), we have $F(t_n) \cap U \neq \emptyset$ for almost every n, and therefore

$$\overline{u}(t_n) = \max_{s\in F(t_n)} u(t_n,s) \geq u(t_o,s_1) \qquad \text{for almost every } n \in \mathbb{N} .$$

Since $\varepsilon>0$ was arbitrary, this shows

$$\underline{\lim}\ \bar{u}(t_n) \geq u(t_o,s_1) = \bar{u}(t_o) .$$

Continuity of $\bar{u}$ follows now immediately from this estimate together with (ii).

Upper semicontinuity of G is a consequence of the continuity of $\bar{u}$ and A.2.3((1) $\Longleftrightarrow$ (2)). □

Finally, some results on measurable correspondences will be cited. Therefore, we assume in the sequel to be given a probability space $(\Omega,\mathcal{A},P)$. A correspondence $F : \Omega \to \mathcal{C}(S)$ for which

$$F^{-1}(\mathcal{S}) \in \mathcal{A} \qquad (\mathcal{S} \in \mathcal{B}(\mathcal{C}(S)))$$

holds, is called an <u>$(\mathcal{A}-\mathcal{B}(\mathcal{C}(S)))$-measurable</u> correspondence. First, we observe the following fact:

<u>A.2.5 Lemma:</u> *Let $F : \Omega \to \mathcal{C}(S)$ be a measurable correspondence. Then, the mapping $d_F : (\omega,s) \mapsto d(s,F(\omega))$ is $\mathcal{A}\otimes\mathcal{B}(S)$-measurable.*

Proof: We first verify that

$$|d(s,C) - d(s',C')| \leq d(s,s') + d^H(C,C')$$

$$(s,s' \in S;\ C,C' \in \mathcal{C}(S)) .$$

Hence, the mapping $(s,C) \mapsto d(s,C)$ is continuous, and therefore $\mathcal{B}(S)\otimes\mathcal{B}(\mathcal{C}(S))$-measurable, since $\mathcal{B}(S\times\mathcal{C}(S)) = \mathcal{B}(S)\otimes\mathcal{B}(\mathcal{C}(S))$. The statement is now a consequence of the $\mathcal{A}\otimes\mathcal{B}(S)-\mathcal{B}(S)\otimes\mathcal{B}(\mathcal{C}(S))$-measurability of the mapping $(\omega,s) \mapsto (s,F(\omega))$. □

The following result will be used to prove the next theorem.

<u>A.2.6 Lemma:</u> *For* $U \subset S$ *let*

$$\mathcal{D}(U) := \{C \in \mathcal{C}(S) \mid C \cap U \neq \emptyset\}$$

and

$$\mathcal{D}'(U) := \{C \in \mathcal{C}(S) \mid C \subset U\} .$$

Then $\mathcal{B}(\mathcal{C}(S))$ *is generated by both systems*

$$\mathcal{S} := \{\mathcal{D}(U) \mid S \supset U \text{ open}\}$$

and

$$\mathcal{S}' := \{\mathcal{D}'(U) \mid S \supset U \text{ open}\} .$$

A proof of this lemma was given by Debreu (1967) .

<u>A.2.7 Theorem:</u> *Let* $\mathcal{W}(S)$ *be the set of probability measures on* $\mathcal{B}(S)$. *We endow* $\mathcal{W}(S)$ *with the weak topology as introduced in A.1 . Then, the mapping*

$$\text{spt} : P \longmapsto \text{spt } P, \quad \mathcal{W}(S) \to \mathcal{C}(S)$$

is $\mathcal{B}(\mathcal{W}(S))$-$\mathcal{B}(\mathcal{C}(S))$*-measurable.*

Proof: By the preceding lemma (A.2.6) we need only verify the relation

$$\text{spt}^{-1}(\mathcal{D}(U)) \in \mathcal{B}(\mathcal{W}(S)) \qquad (U \subset S, U \text{ open}) .$$

To this end let be given an open set $U \subset S$. Then for every $P \in \mathcal{W}(S)$ we have

$$\text{spt } P \in \mathcal{D}(U) \iff \text{spt } P \cap U \neq \emptyset \iff$$
$$\iff \text{spt } P \not\subset S - U \iff P(S-U) < 1 ,$$

where the last equivalence statement results from the compactness of $S - U$. Hence,

$$\text{(i)} \qquad \text{spt}^{-1}(\mathcal{D}(U)) = \{P \in \mathcal{W}(S) \mid P(U) > 0\}$$

holds. By the Portmenteau Theorem (A.1.1), the mapping $P \mapsto P(U)$ is lower semicontinuous, since U is an open set. Therefore and because of (i), $\mathrm{spt}^{-1}(\mathcal{D}(U))$ is an open subset of $\mathcal{W}(S)$, hence contained in $\mathcal{B}(\mathcal{W}(S))$. □

<u>A.2.8 Corollary:</u> *Let* $K : \Omega\times\mathcal{B}(S) \to [0,1]$ *be a stochastic kernel. Then, the mapping*

$$\omega \mapsto \mathrm{spt}\, K(\omega,.), \quad \Omega \to \mathcal{C}(S)$$

is $\mathcal{A}$-$\mathcal{B}(\mathcal{C}(S))$*-measurable.*

Proof: As a consequence of A.1.11, the mapping $\omega \mapsto K(\omega,.)$ is $\mathcal{A}$-$\mathcal{B}(\mathcal{W}(S))$-measurable. The statement is therefore proved by A.2.7. □

The next result is a consequence of A.2.5.

<u>A.2.9 Lemma:</u> *For every measurable correspondence* $F : \Omega \to \mathcal{C}(S)$, *we have*

$$\mathrm{Graph}(F) \in \mathcal{A}\otimes\mathcal{B}(S) .$$

Proof: The compactness of $F(\omega)$ $(\omega\in\Omega)$ implies

$$\mathrm{Graph}(F) = \{(\omega,s)\in\Omega\times S \mid s \in F(\omega)\} =$$

$$= \{(\omega,s)\in\Omega\times S \mid d_F(\omega,s) = 0\} ,$$

when d_F is the function from A.2.5. By A.2.5, d_F is $\mathcal{A}\otimes\mathcal{B}(S)$-measurable. This proves the lemma. □

The following theorem is borrowed from Hildenbrand (1974), Theorem 1, p. 54 (cp. also Sainte-Beuve (1974)).

<u>A.2.10 Theorem:</u> *Every correspondence* $F : \Omega \to \mathcal{B}(S)$ *with*

$$\text{Graph}(F) \in \mathcal{A} \otimes \mathcal{B}(S)$$

allows a measurable selection, i.e., there exists an $\mathcal{A}$-$\mathcal{B}(S)$*-measurable mapping* $f : \Omega \to S$ *with*

$$f(\omega) \in F(\omega) \quad P\text{-a.e.}$$

(It even suffices to assume S *to be a Polish space.)*

A stronger variant of this theorem is known for measurable correspondences $F : \Omega \to \mathcal{C}(S)$ (cp. Hildenbrand (1974), Lemma 1, p. 55).

<u>A.2.11 Lemma:</u> *Let* $F : \Omega \to \mathcal{C}(S)$ *be a measurable correspondence. Then, an* $\mathcal{A}$-$\mathcal{B}(S)$*-measurable mapping* $f : \Omega \to S$ *with*

$$f(\omega) \in F(\omega) \quad (\omega \in \Omega)$$

exists.

Sketch of proof: Since $\mathcal{C}(S)$ is a compact metric space w.r.t. the Hausdorff distance d^H by A.2.1 , we can construct a sequence (F_n) of $\mathcal{A}$-$\mathcal{B}(\mathcal{C}(S))$-measurable correspondences $F_n : \Omega \to \mathcal{C}(S)$ taking only a finite number of values, and having the property

$$d^H(F_n(\omega), F(\omega)) \leq 2^{-n} \quad (n \in \mathbb{N}, \omega \in \Omega) .$$

For every correspondence F_n we find a measurable function $f_n : \Omega \to S$ taking only finitely many values, and satisfying

$$f_n(\omega) \in F_n(\omega) \quad (n \in \mathbb{N}, \omega \in \Omega)$$

as well as

$$d(f_n(\omega), f_{n+1}(\omega)) \leq \frac{1}{2^{n-1}} \quad (n \in \mathbb{N}, \omega \in \Omega) .$$

Thus, $(f_n(\omega))$ becomes a Cauchy sequence which has a limit point $f(\omega)$ in S for every $\omega \in \Omega$. The function f proves to satisfy all requirements. □

A 3 Convex Analysis

The following theorem can be found in Nikaido (1968) as Theorem 2.4, p. 19.

A.3.1 Theorem: *The convex hull* conv A *of an arbitrary set* $A \subset \mathbb{R}^n$ *coincides with the set of all points* $x \in \mathbb{R}^n$ *which have a representation*

$$x = \sum_{i=1}^{n+1} \alpha_i x_i$$

for some $x_1, \ldots, x_{n+1} \in A$ *and* $\alpha_1, \ldots, \alpha_{n+1} \in \mathbb{R}_+$ *with* $\sum_{i=1}^{n+1} \alpha_i = 1$.

The next theorem expresses a result of Dvoretzky, Wald and Wolfowitz (1951), Theorem 2.1 .

A.3.2 Theorem: *Let* $(\Omega, \mathcal{A})$ *be a measurable space, and let* $(\mu_1, \ldots, \mu_q)$ *be a family of finite nonatomic measures on* $\mathcal{A}$. *Further, let* S *be a finite set, and let* $(\alpha_s | s \in S)$ *be a family of measurable functions* $\alpha_s : \Omega \to \mathbb{R}_+$ *with* $\sum_{s \in S} \alpha_s = 1$. *Then, a partition* $(D_s | s \in S)$ *of* Ω *into measurable sets* D_s *exists with*

$$\int \alpha_s \, d\mu_k = \mu_k(D_s) \quad (s \in S;\ k=1,\ldots,q) .$$

The following result is borrowed from Pfanzagl (1974). A proof is given in Ferguson (1967) (Lemma 3 on p. 74).

A.3.3 Lemma: *Let* C *be a convex subset of* $\mathbb{R}^n$, *and let* Q *be a probability measure on* $\mathcal{B}(\mathbb{R}^n)$ *such that the outer measure* $Q^*(C)$ *equals* 1. *Let* $\pi_1, \ldots, \pi_n$ *be the projections of* $\mathbb{R}^n$ *onto its coordinate axises, and let the functions* $\pi_1, \ldots, \pi_n$ *be Q-inte-*

grable. Then,

$$\left(\int \pi_1 \, dQ, \ldots, \int \pi_n \, dQ\right) \in C$$

holds true.

This lemma implies the following fact:

<u>A.3.4 Corollary:</u> *Let* (Ω, A, P) *be a probability space, and let* $f : \Omega \to \mathbb{R}^n$ *be a P-integrable function. Then*

$$\int f \, dP \in \operatorname{conv}\{f(\omega) \mid \omega \in \Omega\}$$

holds true.

Proof: Let Q be the image of P w.r.t. the mapping f ; and let

$$C := \operatorname{conv}\{f(\omega) \mid \omega \in \Omega\}.$$

Obviously, $f(\omega) \in C$ $(\omega \in \Omega)$ holds true. Therefore the outer measure $Q^*(C)$ equals 1. By the transformation theorem for integrals, we have

$$\int f_i \, dP = \int \pi_i \, dQ \quad (i=1,\ldots,n)$$

for $f =: (f_1,\ldots,f_n)$. The statement is now an easy application of A.3.3 . □

Jensen's Inequality is an immediate consequence of A.3.4 .

<u>A.3.5 Corollary:</u> *Let C be a convex subset of* $\mathbb{R}^n$, *and let* $\varphi : C \to \mathbb{R}$ *be a measurable convex function. Further, let* (Ω, A, P) *be a probability space, and let* $f : \Omega \to \mathbb{R}^n$ *be a P-integrable function such that*

$$f(\omega) \in C \quad (\omega \in \Omega).$$

Then, in case of P-*integrability of* $\varphi\circ f$, *the inequality*

$$\varphi\left(\int f\, dP\right) \leqq \int \varphi\circ f\, dP$$

exists.

Proof: Define the set $D \subset \mathbb{R}^{n+1}$ by

$$D := \{(x,t) \in C\times\mathbb{R} \mid \varphi(x) \leqq t\} .$$

Then, D is a convex subset of $\mathbb{R}^{n+1}$ with

$$(f(\omega),\varphi\circ f(\omega)) \in D \quad (\omega\in\Omega) .$$

Let Q be the image of P w.r.t. the mapping $(f,\varphi\circ f)$. Then, the outer measure $Q^*(D)$ equals 1. From A.3.4 , we conclude that

$$\left(\int f\, dP, \int \varphi\circ f\, dP\right) \in D .$$

The desired inequality results now from the definition of the set D. □

A 4 Functions and measures on the unit sphere

Let

$$E := \{z\in\mathbb{C} \mid |z|=1\} = \{e^{i2\pi\varphi} \mid 0\leq\varphi<1\}$$

be the unit sphere of $\mathbb{C}$. Further, let $\mathcal{B}(E)$ be the Borel σ-field over E, when E is endowed with the usual topology generated by the Euclidean metric. Let the uniform distribution λ on $\mathcal{B}(E)$ be the image of the Lebesgue-Borel-measure $\lambda_{[0,1]}$ on $\mathcal{B}([0,1])$ w.r.t. the mapping $\varphi \mapsto e^{i2\pi\varphi}$. In particular, a probability measure μ on $\mathcal{B}(E)$ coincides with λ, if and only if

$$\mu(B_\varphi(z)) = \mu(B_\varphi(z')) \qquad (z,z'\in E,\ \varphi\in[0,2\pi)),$$

where

$$B_\varphi(z) := \{ze^{i\delta} \mid 0\leq\delta<\varphi\} \qquad (z\in E,\ \varphi\in[0,2\pi))$$

is the halfopen arc covered by the point z when it is turned to the left by the angle φ.

A.4.1 Definition: *Let $f : E \to \mathbb{R}$ be a function, and fix $z_o \in E$. Whenever*

$$\lim_{\substack{\varphi\to 0\\ \varphi>0}} \frac{f(z_o e^{i\varphi}) - f(z_o)}{\varphi} =: D_1 f(z_o)$$

exists, we say that f is lefthanded differentiable at z_o.

Whenever f is lefthanded differentiable at every point $z_o \in E$, we say that f is lefthanded differentiable.

A.4.2 Theorem: *Let μ be a probability measure on $\mathcal{B}(E)$, and let $f : E \to \mathbb{R}$ be a lefthanded differentiable function such that*

$$\left| \frac{f(ze^{i\varphi}) - f(z)}{\varphi} \right| \leq c \quad (z\in E,\ \varphi\in (0,\varphi_0))$$

holds for some $c \in \mathbb{R}$ *and* $\varphi_0 > 0$, *and such that*

$$\int f(zz')\, d\mu(z) = \int f(z)\, d\mu(z) \quad (z'\in E) .$$

Then, $D_1 f$ *is bounded and satisfies*

$$\int D_1 f(zz')\, d\mu(z) = 0 \quad (z'\in E) .$$

Proof: For

$$d_n(z) := n\,(f(ze^{i/n}) - f(z)) \quad (z\in E,\ n\in \mathbb{N})$$

we have $|d_n| \leq c$ for almost every $n \in \mathbb{N}$

and

$$\lim_{n\to\infty} d_n = D_1 f .$$

Hence, $D_1 f$ is bounded by c. Moreover, from the Dominated Convergence Theorem, we derive

$$\int D_1 f(zz')\, d\mu(z) = \lim_{n\to\infty} \int d_n(zz')\, d\mu(z) =$$

$$= \lim_{n\to\infty} n \int (f(zz'e^{i/n}) - f(zz'))\, d\mu(z) =$$

$$= \lim_{n\to\infty} n \left(\int f(zz'e^{i/n})\, d\mu(z) - \int f(zz')\, d\mu(z)\right) =$$

$$= 0 . \qquad \square$$

<u>A.4.3 Theorem:</u> *Let be given* $\varphi \in \mathbb{R}$ *with* $0 < \varphi < 2\pi$ *such that* $\varphi/2\pi$ *is irrational. Moreover, let* μ *be a measure on* $\mathcal{B}(E)$ *with* $\mu(E) < \infty$ *satisfying*

$$\mu(B_\varphi(z)) = \mu(B_\varphi(ze^{i\varphi})) \quad (z\in E) .$$

Then, μ *is a multiple of the uniform distribution on* $\mathcal{B}(E)$.

Proof: The proof bases essentially on Kronecker's Theorem (cp. Jacobs (1972), Satz 2.1 , p. 60). This theorem says that the set

$$\{ze^{in\varphi} \mid n \in \mathbb{N}\}$$

is dense in E. Hence, for $z, z' \in E$, a sequence (n_k) of natural numbers exists with $\lim_{k\to\infty} ze^{in_k\varphi} = z'$. The sequence (n_k) can be chosen in such a way that $ze^{in_k\varphi} \notin B_\varphi(z')$ $(k \in \mathbb{N})$ holds. Hence, we have

$$\lim_{k\to\infty} \mu(B_\varphi(ze^{in_k\varphi})) = \mu(B_\varphi(z')) .$$

By our assumptions, we conclude therefrom that

(i) $$\mu(B_\varphi(z)) = \mu(B_\varphi(z'))$$

is satisfied .

Let now $\delta \in (0, 2\pi-\varphi]$ be given. Then, from the identity

$$B_\delta(z) \cup B_\varphi(ze^{i\delta}) = B_\varphi(z) \cup B_\delta(ze^{i\varphi}) \quad (z \in E)$$

we derive

$$\mu(B_\delta(z)) = \mu(B_\delta(ze^{i\varphi})) \quad (z \in E) .$$

The same way of argumentation having led to (i) applies now again and results in

(ii) $$\mu(B_\delta(z)) = \mu(B_\delta(z')) \quad (z, z' \in E) .$$

Since

$$\mathcal{E} := \{B_\delta(z) \mid z \in E,\ \delta \in (0, 2\pi]\}$$

is closed against finite intersections and generates $\mathcal{B}(E)$, the measure μ is rotational-invariant by (ii). The only measures with this property are multiples of the uniform distribution.

□

A 5 Fixed points

The following theorem is a generalization of Kakutani's Fixed Point Theorem, and has been proved by Glicksberg (1952) and Ky Fan (1952).

A.5.1 Theorem: *Let* Σ *be a convex and compact subset of a real locally convex Hausdorff topological vector space, and let* $\Phi : \Sigma \to P(\Sigma)$ *be an upper semicontinuous, convex- and compact-valued correspondence. Then,* Φ *has a fixed point, i.e., there exists* $x \in \Sigma$ *with* $x \in \Phi(x)$.

REFERENCES

Aubin, J.P.; *Mathematical Methods of Game and Economic Theory;* Studies in Math. and its Appl. 7, North Holland, 1979

Aumann, R.J.; Katznelson, Y.; Radner, R.; Rosenthal, R.W.; Weiß, B.; *Approximate purification of mixed strategies;* Math. of Oper. Research, Vol. 8, No. 3, 1983, 327 - 341

Bauer, H.; *Wahrscheinlichkeitstheorie und Grundzüge der Maßtheorie;* W. de Gruyter, 3. Auflage, 1978

Berger, J.O.; *Statistical decision theory and Bayesian analysis;* Springer, New York, 1980

Billingsley, P.; *Convergence of probability measures*; John Wiley & Sons, 1968

Castaing, Ch.; *Sur les multi-applications mesurables*; R.I.R.O., 1re année, No. 1, 1967, 91 - 126

Debreu, G.; *A social equilibrium existence theorem*; Proc. of the Nat. Ac. of Sciences of the USA 38, 1952

Debreu, G.; *Integration of correspondences*; Proc. of the 5th Berkeley Symp. on Math. Stat. and Prob. 2, Vol. 1, 1967, 351 - 372

Dvoretzky, A.; Wald, A.; Wolfowitz, J.; *Elimination of randomization in certain statistical decision procedures and zero-sum two-person games*; Ann. Math. Stat. 22, 1951, 1 - 21

Eberl, W.; Moeschlin, O.; *Mathematische Statistik*; W. de Gruyter, 1982

Ferguson, T.S.; *Mathematical Statistics. A Decision Theoretic Approach*; Academic Press, 1967

Glicksberg, I.L.; *A further generalization of the Kakutani fixed point theorem, with application to Nash equilibrium points*; Proc. of the Nat. Ac. of Sciences 38, 1952, 170 - 172

Greenberg, J.; Shitovitz, B.; Wieczorek, A.; *Existence of Equilibria in atomless production economies with price dependent preferences;* CORE Discussion Paper 7730, Université Catholique de Louvain, 1977

Hausdorff, F.; *Set theory;* Chalsea, New York, 1957

Hildenbrand, W.; *Core and equilibria of a large economy*; Princeton University Press, 1974

Himmelberg, C.J.; *Measurable relations*; Fund. Math. LXXXVII, 1975, 53 - 72

Jacobs, K.; *Selecta Mathematica IV*; Heidelberger Taschenbücher Band 98, Springer Verlag, 1972

Kelley, J.L.; *General Topology*; Van Nostrand, 1961

Kuratowski, K.; *Topology, Vol. II*; Academic Press, 1968

Ky Fan; *Fixed-point and minimax theorems in locally convex topological linear spaces*; Proc. of the Nat. Ac. of Sciences 38, 1952, 121 - 126

Meister, H.; *Ein Bayesscher Risikoansatz für Bereichsschätzfunktionen*; Seminarberichte aus dem Fachbereich Mathematik und Informatik, Fernuniversität Hagen, 1983, 171 - 182

Mertens, J.-F.; Zamir, S.; *Formulation of Bayesian Analysis for Games with Incomplete Information*; Int. Journal of Game Theory, Vol. 14, Issue 1, 1 - 29

Milgrom, P.; Weber, R.; *Distributional strategies for games with incomplete information*; Math. of Oper. Research, Vol. 10, No. 4, 1985, 619 - 632

Moeschlin, O.; *A Bayes concept of confidence estimation;* Transactions of the ninth Prague Conference on Information Theory, Stat. Dec. Functions, Random Processes, Prague 1983

Nash, J.; *Non-cooperative games*; Ann. of Math. 54, 1951, 286 - 295

Nikaido, H.; *Convex Structures and Economic Theory*; Math. in Science and Engineering vol. 51, Academic Press, 1968

Nowak, A.S.; *Measurable selection theorems for minimax stochastic optimization problems*; SIAM J. Control and Optimization, Vol. 23, No. 3, 1985, 466 - 476

Parthasarathy, K.R.; *Probability measures on metric spaces;* Probability and Math. Stat., Academic Press 1967

Pfanzagl, J.; *Convexity and conditional expectations;* Ann. of Prob. 2, 1974, 490 - 494

Radner, R.; Rosenthal, R.W.; *Private information and pure-strategy equilibria;* Math. of Oper. Research, Vol. 7, No. 3, 1982, 401 - 409

Rieder, U.; *Bayesian dynamic programming;* Adv. Appl. Prob. 7, 1975, 330 - 348

Rohatgi, V.K.; *An introduction to probability theory and mathematical statistics;* John Wiley, New York, 1976

Rosenmüller, J.; *Kooperative Spiele und Märkte*; Lecture Notes in Operations Research and Math. Systems 53, Springer Verlag, 1971

Sainte-Beuve, M.-F.; *On the extension of von Neumann-Aumann's Theorem*; Journal of Functional An. 17, 1974, 112 - 129

Shafer, W.; Sonnenschein, H.; *Equilibrium in abstract economies without ordered preferences*; Journ. of Math. Ec. 2, 1975, 345 - 348

Vind, K.; *Edgeworth-allocations in an exchange economy with many traders*; International Ec. Review 5, 1964, 165 - 177

Wald, A.; *Statistical Decision Functions*; Chealsea Publ. Comp., 1971

Wieczorek, A.; *Constrained and indefinite games and their applications*; Institute of Computer Science, Polish Ac. of Sciences, 1984

Vol. 211: P. van den Heuvel, The Stability of a Macroeconomic System with Quantity Constraints. VII, 169 pages. 1983.

Vol. 212: R. Sato and T. Nôno, Invariance Principles and the Structure of Technology. V, 94 pages. 1983.

Vol. 213: Aspiration Levels in Bargaining and Economic Decision Making. Proceedings, 1982. Edited by R. Tietz. VIII, 406 pages. 1983.

Vol. 214: M. Faber, H. Niemes und G. Stephan, Entropie, Umweltschutz und Rohstoffverbrauch. IX, 181 Seiten. 1983.

Vol. 215: Semi-Infinite Programming and Applications. Proceedings, 1981. Edited by A.V. Fiacco and K.O. Kortanek. XI, 322 pages. 1983.

Vol. 216: H.H. Müller, Fiscal Policies in a General Equilibrium Model with Persistent Unemployment. VI, 92 pages. 1983.

Vol. 217: Ch. Grootaert, The Relation Between Final Demand and Income Distribution. XIV, 105 pages. 1983.

Vol. 218: P. van Loon, A Dynamic Theory of the Firm: Production, Finance and Investment. VII, 191 pages. 1983.

Vol. 219: E. van Damme, Refinements of the Nash Equilibrium Concept. VI, 151 pages. 1983.

Vol. 220: M. Aoki, Notes on Economic Time Series Analysis: System Theoretic Perspectives. IX, 249 pages. 1983.

Vol. 221: S. Nakamura, An Inter-Industry Translog Model of Prices and Technical Change for the West German Economy. XIV, 290 pages. 1984.

Vol. 222: P. Meier, Energy Systems Analysis for Developing Countries. VI, 344 pages. 1984.

Vol. 223: W. Trockel, Market Demand. VIII, 205 pages. 1984.

Vol. 224: M. Kiy, Ein disaggregiertes Prognosesystem für die Bundesrepublik Deutschland. XVIII, 276 Seiten. 1984.

Vol. 225: T.R. von Ungern-Sternberg, Zur Analyse von Märkten mit unvollständiger Nachfragerinformation. IX, 125 Seiten. 1984

Vol. 226: Selected Topics in Operations Research and Mathematical Economics. Proceedings, 1983. Edited by G. Hammer and D. Pallaschke. IX, 478 pages. 1984.

Vol. 227: Risk and Capital. Proceedings, 1983. Edited by G. Bamberg and K. Spremann. VII, 306 pages. 1984.

Vol. 228: Nonlinear Models of Fluctuating Growth. Proceedings, 1983. Edited by R.M. Goodwin, M. Krüger and A. Vercelli. XVII, 277 pages. 1984.

Vol. 229: Interactive Decision Analysis. Proceedings, 1983. Edited by M. Grauer and A.P. Wierzbicki. VIII, 269 pages. 1984.

Vol. 230: Macro-Economic Planning with Conflicting Goals. Proceedings, 1982. Edited by M. Despontin, P. Nijkamp and J. Spronk. VI, 297 pages. 1984.

Vol. 231: G.F. Newell, The M/M/∞ Service System with Ranked Servers in Heavy Traffic. XI, 126 pages. 1984.

Vol. 232: L. Bauwens, Bayesian Full Information Analysis of Simultaneous Equation Models Using Integration by Monte Carlo. VI, 114 pages. 1984.

Vol. 233: G. Wagenhals, The World Copper Market. XI, 190 pages. 1984.

Vol. 234: B.C. Eaves, A Course in Triangulations for Solving Equations with Deformations. III, 302 pages. 1984.

Vol. 235: Stochastic Models in Reliability Theory. Proceedings, 1984. Edited by S. Osaki and Y. Hatoyama. VII, 212 pages. 1984.

Vol. 236: G. Gandolfo, P.C. Padoan, A Disequilibrium Model of Real and Financial Accumulation in an Open Economy. VI, 172 pages. 1984.

Vol. 237: Misspecification Analysis. Proceedings, 1983. Edited by T.K. Dijkstra. V, 129 pages. 1984.

Vol. 238: W. Domschke, A. Drexl, Location and Layout Planning. IV, 134 pages. 1985.

Vol. 239: Microeconomic Models of Housing Markets. Edited by K. Stahl. VII, 197 pages. 1985.

Vol. 240: Contributions to Operations Research. Proceedings, 1984. Edited by K. Neumann and D. Pallaschke. V, 190 pages. 1985.

Vol. 241: U. Wittmann, Das Konzept rationaler Preiserwartungen. XI, 310 Seiten. 1985.

Vol. 242: Decision Making with Multiple Objectives. Proceedings, 1984. Edited by Y.Y. Haimes and V. Chankong. XI, 571 pages. 1985.

Vol. 243: Integer Programming and Related Areas. A Classified Bibliography 1981–1984. Edited by R. von Randow. XX, 386 pages. 1985.

Vol. 244: Advances in Equilibrium Theory. Proceedings, 1984. Edited by C.D. Aliprantis, O. Burkinshaw and N.J. Rothman. II, 235 pages. 1985.

Vol. 245: J.E.M. Wilhelm, Arbitrage Theory. VII, 114 pages. 1985.

Vol. 246: P.W. Otter, Dynamic Feature Space Modelling, Filtering and Self-Tuning Control of Stochastic Systems. XIV, 177 pages. 1985.

Vol. 247: Optimization and Discrete Choice in Urban Systems. Proceedings, 1983. Edited by B.G. Hutchinson, P. Nijkamp and M. Batty. VI, 371 pages. 1985.

Vol. 248: Plural Rationality and Interactive Decision Processes. Proceedings, 1984. Edited by M. Grauer, M. Thompson and A.P. Wierzbicki. VI, 354 pages. 1985.

Vol. 249: Spatial Price Equilibrium: Advances in Theory, Computation and Application. Proceedings, 1984. Edited by P. T. Harker. VII, 277 pages. 1985.

Vol. 250: M. Roubens, Ph. Vincke, Preference Modelling. VIII, 94 pages. 1985.

Vol. 251: Input-Output Modeling. Proceedings, 1984. Edited by A. Smyshlyaev. VI, 261 pages. 1985.

Vol. 252: A. Birolini, On the Use of Stochastic Processes in Modeling Reliability Problems. VI, 105 pages. 1985.

Vol. 253: C. Withagen, Economic Theory and International Trade in Natural Exhaustible Resources. VI, 172 pages. 1985.

Vol. 254: S. Müller, Arbitrage Pricing of Contingent Claims. VIII, 151 pages. 1985.

Vol. 255: Nondifferentiable Optimization: Motivations and Applications. Proceedings, 1984. Edited by V.F. Demyanov and D. Pallaschke. VI, 350 pages. 1985.

Vol. 256: Convexity and Duality in Optimization. Proceedings, 1984. Edited by J. Ponstein. V, 142 pages. 1985.

Vol. 257: Dynamics of Macrosystems. Proceedings, 1984. Edited by J.-P. Aubin, D. Saari and K. Sigmund. VI, 280 pages. 1985.

Vol. 258: H. Funke, Eine allgemeine Theorie der Polypol- und Oligopolpreisbildung. III, 237 pages. 1985.

Vol. 259: Infinite Programming. Proceedings, 1984. Edited by E.J. Anderson and A.B. Philpott. XIV, 244 pages. 1985.

Vol. 260: H.-J. Kruse, Degeneracy Graphs and the Neighbourhood Problem. VIII, 128 pages. 1986.

Vol. 261: Th.R. Gulledge, Jr., N.K. Womer, The Economics of Made-to-Order Production. VI, 134 pages. 1986.

Vol. 262: H.U. Buhl, A Neo-Classical Theory of Distribution and Wealth. V, 146 pages. 1986.

Vol. 263: M. Schäfer, Resource Extraction and Market Structure. XI, 154 pages. 1986.

Vol. 264: Models of Economic Dynamics. Proceedings, 1983. Edited by H.F. Sonnenschein. VII, 212 pages. 1986.

Vol. 265: Dynamic Games and Applications in Economics. Edited by T. Başar. IX, 288 pages. 1986.

Vol. 266: Multi-Stage Production Planning and Inventory Control. Edited by S. Axsäter, Ch. Schneeweiss and E. Silver. V, 264 pages. 1986.

Vol. 267: R. Bemelmans, The Capacity Aspect of Inventories. IX, 165 pages. 1986.

Vol. 268: V. Firchau, Information Evaluation in Capital Markets. VII, 103 pages. 1986.

Vol. 269: A. Borglin, H. Keiding, Optimality in Infinite Horizon Economies. VI, 180 pages. 1986.

Vol. 270: Technological Change, Employment and Spatial Dynamics. Proceedings 1985. Edited by P. Nijkamp. VII, 466 pages. 1986.

Vol. 271: C. Hildreth, The Cowles Commission in Chicago, 1939–1955. V, 176 pages. 1986.

Vol. 272: G. Clemenz, Credit Markets with Asymmetric Information. VIII, 212 pages. 1986.

Vol. 273: Large-Scale Modelling and Interactive Decision Analysis. Proceedings, 1985. Edited by G. Fandel, M. Grauer, A. Kurzhanski and A.P. Wierzbicki. VII, 363 pages. 1986.

Vol. 274: W.K. Klein Haneveld, Duality in Stochastic Linear and Dynamic Programming. VII, 295 pages. 1986.

Vol. 275: Competition, Instability, and Nonlinear Cycles. Proceedings, 1985. Edited by W. Semmler. XII, 340 pages. 1986.

Vol. 276: M.R. Baye, D.A. Black, Consumer Behavior, Cost of Living Measures, and the Income Tax. VII, 119 pages. 1986.

Vol. 277: Studies in Austrian Capital Theory, Investment and Time. Edited by M. Faber. VI, 317 pages. 1986.

Vol. 278: W.E. Diewert, The Measurement of the Economic Benefits of Infrastructure Services. V, 202 pages. 1986.

Vol. 279: H.-J. Büttler, G. Frei and B. Schips, Estimation of Disequilibrium Models. VI, 114 pages. 1986.

Vol. 280: H.T. Lau, Combinatorial Heuristic Algorithms with FORTRAN. VII, 126 pages. 1986.

Vol. 281: Ch.-L. Hwang, M.-J. Lin, Group Decision Making under Multiple Criteria. XI, 400 pages. 1987.

Vol. 282: K. Schittkowski, More Test Examples for Nonlinear Programming Codes. V, 261 pages. 1987.

Vol. 283: G. Gabisch, H.-W. Lorenz, Business Cycle Theory. VII, 229 pages. 1987.

Vol. 284: H. Lütkepohl, Forecasting Aggregated Vector ARMA Processes. X, 323 pages. 1987.

Vol. 285: Toward Interactive and Intelligent Decision Support Systems. Volume 1. Proceedings, 1986. Edited by Y. Sawaragi, K. Inoue and H. Nakayama. XII, 445 pages. 1987.

Vol. 286: Toward Interactive and Intelligent Decision Support Systems. Volume 2. Proceedings, 1986. Edited by Y. Sawaragi, K. Inoue and H. Nakayama. XII, 450 pages. 1987.

Vol. 287: Dynamical Systems. Proceedings, 1985. Edited by A.B. Kurzhanski and K. Sigmund. VI, 215 pages. 1987.

Vol. 288: G.D. Rudebusch, The Estimation of Macroeconomic Disequilibrium Models with Regime Classification Information. VII, 128 pages. 1987.

Vol. 289: B.R. Meijboom, Planning in Decentralized Firms. X, 168 pages. 1987.

Vol. 290: D.A. Carlson, A. Haurie, Infinite Horizon Optimal Control. XI, 254 pages. 1987.

Vol. 291: N. Takahashi, Design of Adaptive Organizations. VI, 140 pages. 1987.

Vol. 292: I. Tchijov, L. Tomaszewicz (Eds.), Input-Output Modeling. Proceedings, 1985. VI, 195 pages. 1987.

Vol. 293: D. Batten, J. Casti, B. Johansson (Eds.), Economic Evolution and Structural Adjustment. Proceedings, 1985. VI, 382 pages. 1987.

Vol. 294: J. Jahn, W. Krabs (Eds.), Recent Advances and Historical Development of Vector Optimization. VII, 405 pages. 1987.

Vol. 295: H. Meister, The Purification Problem for Constrained Games with Incomplete Information. X, 127 pages. 1987.